Civilizi

Civilizing Natures

Race, Resources, and Modernity in Colonial South India

Kavita Philip

Rutgers University Press
New Brunswick, New Jersey

Library of Congress Cataloging-in-Publication Data

Philip, Kavita, 1964–
Civilizing natures : race, resources, and modernity in colonial South India / Kavita Philip.
p. cm.
Includes bibliographical references and index.
ISBN 0–8135–3360–0 (hardcover: alk. paper) — ISBN 0–8135–3361–9 (pbk. : alk. paper)
1. Science—India, South—History—19th century. 2. Science—Social aspects—India, South—History—19th century. I. Title.
Q127.I4P48 2004
509.548'09'034—dc21

2003005675

British Cataloging-in-Publication information is available from the British Library

Manufactured in the United States of America

Contents

Acknowledgments *vii*

1 *Introduction* *1*

2 *A Local Story: English Mud* *29*

3 *Forests* *54*

4 *Plantations* *80*

5 *Ethnographers* *99*

6 *Christianity* *147*

7 *A Global Story: Imperial Science Rescues a Tree* *171*

8 *Conclusion* *196*

Notes *203*

Index *237*

Acknowledgments

For enabling the beginnings of this project, thanks to Peter Taylor, Ron Kline, Satya Mohanty, and Bernard Cohn, who first encouraged me to follow all my lines of interest, and never squelched eccentric interdisciplinary predilections. Special thanks to Barney for teaching me how to read the archives, and for his unwavering enthusiasm for the project; and to Ron, who urged me to tell the interesting stories when the archival data looked endless and unwieldy. Sander Gilman showed me how histories of science can be rigorous, timely, and relevant; Dick Boyd and Sergio Sismondo helped me understand how history might elucidate questions from the philosophy of science. Dean Sue Rosser and my LCC chairs helped me find writing time at a critical juncture. I am grateful to the anonymous reviewers of the manuscript, and to staff at the India Office Library and at the University of Chicago's Regenstein Library. I am enormously fortunate to have worked with Rutgers University Press's scholar-editors: acquisition editor and historian of science Audra Wolfe, whose erudition, patience, and efficiency were astonishing; and Margaret Case, who combines matchless copyediting skills with deep scholarship in South Asian history.

Thanks to Kim Berry, Mark Baker, Priyamvada Gopal, Shubhra Gururani, and David Takacs for inspiring me with their work; and for sharing scholarly insights, side-splitting humor, and bountiful kitchens with me over the many years this project has lived and grown. And for showing me how to love writing history while embracing all the worlds we live and work in today, endless and ongoing thanks to Ward Smith.

For permission to reprint material from earlier versions of Chapters 2 and 7, I thank the journals *Cultural Studies* (available online at *http://*

www.tandf.co.uk) and *Environment and History* (online at *http://www.erica.demon.co.uk/EH.html*). These chapters originally appeared as "English Mud: Towards a Critical Cultural Studies of Colonial Science" in *Cultural Studies* 12.3 (1998): 300–331, and "Imperial Science Rescues a Tree: Global Botanic Networks, Local Knowledge, and the Transcontinental Transplantation of Cinchona" in *Environment and History* 1.2 (1995): 173–200.

For permission to reproduce images, I thank the following: Asian Educational Services, for permission to reproduce several plates from Edgar Thurston assisted by K. Rangachari, *Castes and Tribes of Southern India*, vols. 1 and 3 (1909; reprint New Delhi: Asian Educational Services, 1987); and the British Library, for permission to reproduce "Lord and Lady Curzon at a tiger hunt," "Members of the Toda tribe," and "Photographs of Andaman Islanders," from Ray Desmond, *Victorian India in Focus* (London: Her Majesty's Stationery Office, 1982). The photograph I took on the grounds of Kew is reproduced with the kind permission of the Board of Trustees, Royal Botanic Gardens, Kew.

Civilizing Natures

Chapter 1

Introduction

New, yet deeply familiar, post- and neocolonial forms of globalization are altering the political economy of resource use on the South Asian subcontinent today. Even as we begin to digest the insights of historians into the colonial period and the implications of historical formations for our contemporary beliefs and actions, the patterns of modernity are being reshaped by this most recent phase of globalization. The new globalization bears many similarities to the old, in terms of the global dimensions of the networks of commercial, political, and cultural power. Economic historians from Immanuel Wallerstein to Paul Hirst and Grahame Thompson have suggested that capitalism has been a global system at least since the sixteenth century. Telegraph cables, steel-hulled ships, and multinational manufacturing firms opened up every continent to global capital flows by the mid-nineteenth century, with innovations as dramatic in scope and scale as those we see in today's communications revolution.[1] Hirst and Thompson comment: "If the theorists of globalization mean that we have an economy in which each part of the world is linked by markets sharing close to real-time information, then that began not in the 1970s but in the 1870s."[2] The history of nineteenth- and twentieth-century science illustrates not only the dramatic shaping of still-evolving global networks of theory and practice but also the forging of a scientific and technological worldview that continues to shape postcolonial natures and cultures. This book interrogates nineteenth- and early twentieth-century practices of science in British colonial India, seeking to track the ways in which scientific modernity functioned at the level of everyday experience, and the modes by which it rendered commonsensical—and thus remarkably persistent—its historically and culturally particular modes of perception and action.

The rhetorical and material traffic among imperial science, culture, and commerce in the nineteenth century had lasting consequences—in both metropole and margin—for the social construction of modernity and the institutional practice of science. Many postcolonial states have carried over colonial bureaucratic and administrative systems, along with the foundational assumptions of scientific modernity, into structures of postindependence government and popular culture. Contemporary social movements in India relating to issues of science, technology, and development range from protest movements around issues such as environmental conservation, equitable access to resources, and industrial safety,[3] to public debates on the legacy of colonial science, the need for national self-sufficiency in technology development, and the role of multinationals in bringing industrial and technological change to the region.[4] To carry out nonreductive analyses of new social movements and to formulate alternatives to conventional narratives of progress, we know we must transcend traditional disciplinary boundaries. But we must do so carefully, so as not to fall into clichéd, revivalist, or essentialist explanations of the effects of scientific modernity.

Scholars in a wide range of disciplines have recently made efforts to bridge disparate methodologies in order to explain ecological and global dynamics. Such multidisciplinary efforts characterize not only the social sciences and humanities but the earth and life sciences too. A synthetic framework of paleontology, biogeochemistry, and botany, for instance, can contribute to the understanding of climate change; or the insights of natural resource management combine with biological research to elucidate the long-term dynamics of ecosystems. The building of synthetic bridges is a long and arduous process that involves the deepening of our disciplinary understandings, rather than superficial cross-border borrowings. This book joins these efforts to synthesize the recent insights of social scientific research. Its methodology is drawn from the humanities and social sciences; it hopes to contribute not only to synthetic efforts in these fields but also, perhaps indirectly, to ongoing efforts to bridge scientific with historical research in environmental studies.[5]

This is not a comprehensive history of science in colonial India; rather, it is an attempt to discern recurring patterns of thought and action that account not only for colonial configurations of power and knowledge but also for the persistence of these configurations in postcolonial representations of race and resources. Such an enterprise, it might be argued, smacks of presentism, a label that, for many a serious reader, delegitimates a properly historical work. Certainly, the questions that shape this work are motivated by current, and to my mind urgent, political and philosophical questions about the power of science to shape the common sense of individuals and the developmental trajectories of states. This book is motivated by the desire to historicize thoroughly our current understandings of the myriad interests aspiring to preserve, use, nurture, protect, or speak on behalf of nature. Such a motivation accounts for the wide and perhaps eclectic choice of foci. The following chapters explore the meanings of labor, progress, productivity, and salvation, and range over the discourses of ethnography, forestry, botany, gardening, plantation management, and Christian proselytization. The wish to historicize current understandings also shapes the selection of events and trends covered in these chapters, and my interpretation of the archival evidence.

The book's argument is, nevertheless, rooted in a very specific regional archival base, and each chapter locates itself within the historical debates around a particular scientific discourse, such as the recently vigorous histories of colonial anthropology and forestry. To many readers, the current theoretical debates in science studies and political philosophy to which I allude may be irrelevant; but it is my hope that the range of archival material presented might persuade the transdisciplinary reader that, whether one's interest is environmentalism, the global economy, or science education, the complexity of modernity demands that we recognize its overlapping influences, effects, and causes in disparate disciplines. I hope a sense of this overlapping complexity will emerge as chapters 2 through 7 track scientific and administrative practices through official and everyday understandings of race, resources, and progress.

Science and Colonialism

Science was a central pillar of colonialism, but the converse holds too: colonialism was central to the history of nineteenth-century science. The history of Western science, particularly as it developed in Britain after the Industrial Revolution, is incomplete without an account of the ways in which science—as an intellectual and philosophical project, and later as an institutionalized and professionalized field—was constituted in relation to the enterprise of colonialism.

Imperialism and the pursuit of objective scientific knowledge were two of the most powerful projects of England after the Industrial Revolution. In the course of the nineteenth century, a mutually sustaining set of practices, institutions, and bureaucratic structures were developed that served both the material and ideological needs of the empire. The twentieth-century notions (which post-independence Nehruvian India embraced) of scientific rationality and of the reasoning, autonomous subject were constituted, to a large extent, in opposition to a notion of the uncivilized, prescientific colonized subject (the latter owing its definition to nineteenth-century anthropological theories of "man"). Twentieth-century constructions of scientific rationality have been determined to a significant extent by historical strategies of deploying science in the colonies, which developed into flexible and subtle yet effective "modern" systems of control and management in reaction to the ostensibly chaotic and irrational forms of resistance they encountered.

Both as a scholarly discipline and as a concept in the popular imagination, science was a critical resource in the achievement of political and cultural hegemony in the British Empire. It was also of central importance in the formulation of counter-hegemonic strategies, which resulted in formulations of progress from both "above" and "below" whose specters continue to haunt visions of postcolonial modernity.

Histories of Colonial Science

Scholarship on the history of colonial science has been particularly productive in the 1980s and 1990s, and has yielded a significant body of archival research on the major forms of colonial science (particularly medicine, forestry, and anthropology) at all the sites of nineteenth-century European (particularly British, French, and Dutch) colonization. The field is beginning to be large and deep enough to reveal a healthy diversity in methodology and ideology—a sign that it has come of age. Within British colonial history alone, we see several strands. One strand of the colonial historiography of British science has given us institutional and intellectual genealogies: we have histories of forestry,[6] hunting,[7] botanical gardens,[8] physiocratic environmentalism, and island ecology.[9] Another strand has surveyed the sciences and technologies that were professionalized and institutionalized in colonial South Asia.[10] Perhaps the most vigorous subfield is environmental history, in which South Asian colonial histories of nature have developed enough to have sparked sustained dialogues with environmental discourses in other colonial regions and periods.[11] Even outside the disciplinary confines of the small field known as history of science, political and cultural historians have noted the central ideological role that science played in shaping imaginations in both Britain and the colonies.[12] Predating—and nurturing—this spurt of historical interest in colonial science were the critiques of colonial scientific ideology that inspired many of us to reexamine our scientific educations: the passionate and polemical work of South Asian public intellectuals such as Shiv Viswanathan, Ashis Nandy, and Vandana Shiva, who continue to shape the critical debate on science.[13]

This extensive scholarship has brought us to the point where it seems appropriate to use it as a springboard for a dialogic interpretation of the past. Each chapter that follows takes on a different slice of the colonial histories of nineteenth- and early twentieth-century science in India, in an attempt to discern the ways in which the patterns of modernity took shape in our institutions and captured our imaginations. Although it is impossible to summarize the state of research in each area, it will be evident that this work builds on, and owes a great deal to, each of the strands above. I have tried to strike a balance between reviewing the history of the debates for the unfamiliar reader and developing the selective analysis of what I found the most compelling aspects of the construction of technoscientific modernity.

Much recent scholarship on the history of science and empire has explored the ways in which science and political culture were mutually constitutive, as were global and local knowledge. In Britain, the rise of science to cultural power was to a large extent dependent on the logic of empire. Whether at home among the working class of London's East End, in rural Ireland, or in India, the jewel of the imperial crown, scientists found that their expertise in the study of nature and the native could chart a route to influence, power, and prestige for themselves as well as for their motherland.[14] The systematic production of knowledge about the world and its inhabitants became not just a by-product of empire but also the fuel by which its engines were run. Britain's power and prosperity depended on the reliable production of information about how to optimize the productivity of people and places distant from its center. The geographic mapping of new territories, the cultivation of economic plants for the productive use of colonial land, agricultural theories of rural progression, anthropological understandings of the worker, the noble savage, and the cultured mind, social-scientific theories ranging from economics to linguistics, became important to the British Crown and Parliament, to adventurers and entrepreneurs, and to colonial administrations. These and other applied sciences are significant to the historian not just as an illustration of how the content of scientific research informed the politics of empire but also because of the forms science began to take as a result of this growing political power. Western science, it turns out, springs from indigenous European as well as indigenous non-Western roots. The institutionalization of these hybrid sciences, and the professionalization of a class of proselytizing scientist-administrators, had lasting effects on the institutional forms of scientific practice as well as the political power of the scientific expert in later eras.[15]

One of the effects of nineteenth-century scientific theories of non-Western nature and natives is the belief, persistent to this day, in an epistemological divide between universal science and local knowledge. Any historical understanding of British science in the nineteenth and early twentieth centuries must contradict such attempts to draw essential distinctions between global and local knowledge. Local knowledges from the peripheries of empire were constitutive of both the form and content of science at the metropolitan center. For instance, indigenous Indian knowledge of forests and cropping techniques influenced the thinking of late-eighteenth-century Hippocratic and Physiocratic thinkers in Britain, spurring the theory and practice of a romanticist environmentalism. Caribbean and South Asian plant species radically changed the shape of European botany in the eighteenth and nineteenth centuries, in terms of taxonomic advances and by altering the geographic scope and economic power of botanical research. Anthropology was dependent on colonial infrastructures—including scientific voyages and botanical projects—in order to obtain access to its object of study, the non-Western native, who in intricately constrained ways participated in the construction of anthropological knowledge. Botany, anthropology, and forestry in the nineteenth century (like many other sciences, from geography to linguistics) had their most important sources of data, their objects of investigation, located at colonial sites. The colonies were the laboratories for nineteenth-century sciences of race and of resources.

Drawing the maps of global power was simultaneous with, and constitutive of, the process of drawing the maps of self and other, nature and culture. Histories of colonial science, then, ought not merely add a history of science in the Third World on to an unaltered history of Western science; they should not serve merely as a consolation prize to now-decolonized natives who, scattered about the Western world, clamor for their own histories. Rather, they entail the critique of our current narratives of the history of Western science itself, and their rewriting into an integrated narrative that analyzes the discourses of exploitation and romanticism within the same theoretical framework that analyzes the cognitive claims of science. The practice of colonial science was not the simple execution of an instrumentalist exploitation of native and nature. Its full complexity includes the dialectic between two forms of post-Enlightenment perception of the jungle and the primitive: as a potential source of material resources to be harnessed to production, and a site to rediscover the romanticist harmonies and aesthetic resources that industrial progress seemed to have dispelled from the mother country's hinterland.

The construction of scientific modernity was not a simple, unidirectional or premeditated strategy; rather, it developed in contradictory ways, demolishing myths of the old order only by constructing spectacular new myths, many of which structure our debates over science today. Nor was modernity a plot perpetrated by a few masterminds of scientific thought. Diverse agents were intertwined in the same intellectual networks of science and the civilizing practices of modernity, including colonial ethnographers, their Brahmin emulators, and their tribal subjects; foresters and their indigenous work force; plantation owners and their laborers; missionaries and their flock.

The postcolonial Indian state's development model grew out of some of the enduring beliefs of the nineteenth century—including not only its myths of absolute scientific objectivity and instrumental efficacy but also its romanticized modes of imagining rebellion. Native subjectivities, and modes of social behavior, were engineered in ways that would enable the success of technoscientific planning for the nation-state. These ways of being are scripted into postcolonial modes of both conservatism and rebellion—modes that require further excavation and analysis. In the following chapters I attempt to sketch the terrain on which such an analysis might begin, and the intellectual currents with which it must contend.

Civilizing Natures' starting point is the recognition that British colonialism and the systematic ordering of objective knowledge were inextricably intertwined. The two were mutually dependent projects; science was not the mere pawn of imperialism nor, conversely, was imperialism the simple reflection of a putatively inherent dominating impulse in scientific thought. Colonial science was not just about controlling the other—it was, simultaneously, about defining the self and its place in nature. And, although the Western male scientist may have been the most powerful agent of science, the parameters of the self were being drawn with the same palette not only for him but also for Western women and non-Western colonial subjects. The postcolonial subject has taken on more aspects of this selfhood than he has rejected. That in itself does not constitute a paradox, for it would be surprising if we found no historical continuities in institutions and psyches. A contradiction arises, however, when this multiply constituted subject dehistoricizes her own sense of self, hoping with a burst of polemic and a dash of wishful imagination to shed her Enlightenment forms of perception and rediscover the authentic indigene within. While it is true that we need critical reevaluations of every aspect of the global networks of economy and fantasy that the last three centuries have washed onto the subcontinent's shores, we cannot simply transcend them by short-circuiting the contradictory dialectic of which our own imaginations are a part.

Understanding nineteenth-century science in its historical context means that we acknowledge, for instance, that penetration and control (the easily critiqued Baconian vices) were as much a part of the scientific spirit as were romanticism and well-intentioned reformism. Much polemical Enlightenment bashing has gone hand-in-hand with the celebration of the apparently aberrant behavior of some "good" Westerners. If the instrumentalist scientist is the villain of the post-colonialist narrative, the romanticist scientist (that rebel against the reductionist scientific mind) is the hero, the friend of the native and of nature, the defender of organic essence against machinic efficiency. So, for example, the *Robinsonnade* naturalists and the physiocratic doctors of the eighteenth and nineteenth centuries, the Scottish biologist and town planner Patrick Geddes and native-loving anthropologist Verrier Elwin have been held up in recent years as encouraging examples of how scientists did travel paths friendly to nature and to natives. But this easy discovery of resistance forgets that romanticism too was the child of the Enlightenment, born out of a binary opposition to its more official face, instrumental rationalism. For every Robinson Crusoe and Prospero striving to comprehend island ecologies, there is always a Man Friday and a Caliban, whose material labors support their masters' metaphysical goals and provide the creature comforts without which romanticist thought does not occur. For every friendly colonial anthropologist, there are the tribal wives whose stories are not yet written.

If we reject the Enlightenment model of science as well as the nostalgic critique that comes with its simple inversion, we lose the awe-inspiring confidence of scientific objectivity (such as that embodied by the technological planners of modern states) as well as comforting romanticist utopias (such as those dear to a wide range of science critics). What's left, one might ask, for those who yearn for a more just order of things? The alternative, of course, is messier than either of the neat poles of the rejected opposition. If we start by unmasking the conditions of production of this binary, we are still left with the search for a synthesis that does not reproduce the blind spots of either pole of the contradiction. Methodologically, one might combine an attention to close readings of scientific rhetoric and ideological critique of specific modernizing discourses with a recognition of the configurations of global power, knowledge, and commerce that framed them.

Cultural Materialist Readings of Nature's History

Colonial constructions of nature, natives, and modernity were mutually constitutive. The field of cultural studies of science has, in the last decade, taken on the task of understanding how ideas of nature have historically shaped and been shaped by social, political, and economic relationships. This calls for attention to processes of representation, the production of ideology, hegemony, the relations between the domination of groups and the production of knowledge about them, and the institutionalization of unequal power relations. Insofar as this book weaves its narrative through all of these issues, it may be seen as a cultural study rather than a traditional history of science. However, it is in many ways materialist cultural and political history with wide-angled perspectives. It tells a story that is situated regionally and temporally, and it assumes that all its protagonists had agency and practical rationality, constrained in complex ways. South India in the period of high imperialism has been thoroughly researched from colonialist, nationalist, and subaltern perspectives. I seek not to revise these narratives wholesale but to unravel many little stories that are often lost in the better-known big stories about nationalism, on the one hand, or about hegemonic scientific theories, on the other. If we pay attention to the nondisciplinary and lay understandings of science, we come to see that modern science draws its power not from any essential violence in its definition, but from the intricate and mundane ways in which it is etched into the daily routines of people, practices, and institutions.[16]

By the end of the twentieth century, radical critiques of technoscientific rationality had been made by the academic discipline of cultural studies, which, in the 1980s and 1990s, called for new (nonlinear, nontotalizing) political philosophies. In this area of scholarship it is often heard that Western science—as a hegemonic set of ideas—was responsible for importing instrumentalist notions of nature into non-Western contexts. Such a claim says both too much and too little.

It says too much because it imputes enormous intentional and performative power to ideas in themselves. The laws of physics were not constructed as instruments for world domination; scientific axioms do not embody the unstoppable juggernaut of hegemony. No matter how much we deconstruct the ideas in themselves, they will not yield up the magic formula for power. The claim that science as a set of ideas, or as a mode of thought, is inherently violent (or, conversely, inherently noble) is undergirded by an idealist fallacy that places too much faith in the explanatory power of pure thought and the essences that reside in its notional realm, unsullied by the messy hand of practice.

It says too little, on the other hand, because it does not take into account the myriad instrumentalist uses of nature in precolonial India, and the active interaction (as opposed to the alleged static harmonies) between groups, castes, kingdoms, and their natural environments. It takes for granted the assumption that science is inherently Western and its obverse (mysticism, traditionalism, or antimodernism) inherently Eastern. It forgets that this putatively eternal dichotomy between West and non-West was constructed recently, by, among other forces, historically specific (not inherently rational or eternally violent) forms of scientific practice. Positivist assertions of the apolitical, value-free truth status of science depend on the belief in a dichotomy between fact and value, by which scientific fact is seen to be superior to traditional belief. A simple inversion of this dichotomy yields the claim that spiritualist tradition is superior to scientific modernity by virtue of its monopoly over the discourse of value. The latter belief often poses as a radical critique of modernity; yet to hold it is to question too little, not too much, of the myths of Western civilization. The claim requires us to assume the truth of mythical binaries whose origins were written by the relatively recent practices of modern capitalism, Cartesianism, and nationalism—not nurtured by ancient traditions. The value-laden oppositions between scientific and religious thought, Western rationality and Eastern spiritualism, instrumental efficacy and meditative contemplation, advancing modernity and static tradition undergird varieties of Enlightenment humanism as well as traditionalist rejections of Enlightenment modernity. It is the complexity of these multiple influences that is lost in attempts to protect the purified essence of the local against the encroachments of the global. Attempts to rediscover a nature outside of history, and a local uncorrupted by the global, are hobbled by a failure to take account of the always-already-mixed cultural histories of nature, and the non-Western histories of Western knowledge. This book offers a cultural materialist reading of some very powerful ideas of nature in a representative corner of the nineteenth-century colonial world, as part of an ongoing academic conversation about nature's history in both European and non-European imaginations.

A cultural materialist understanding of colonial science does not satisfy the political desire, expressed by many cultural theorists and activists, to be able to dismiss all Western scientific thought as contaminated by violence. It leaves us, instead, with more complicated questions about the different modes in which modern knowledge was constructed through encounters among multiple modes of instrumentalist and aestheticist comprehensions of nature and culture. Political-economic and cultural processes offer a more historically contingent picture of scientific rationality, but render unviable the nostalgia for a precolonial innocence. I adopt this cultural materialist approach not in order to reject political critique of science, however, but in an attempt to develop a more complex political engagement between the discipline of the history and philosophy of science, and the political formations of post-colonial scientific and environmental development.

No more viable than a romantic desire for cleansing the technoscientific psyche is the mainstream disciplinary conservatism that continues to carve up historical studies according to geographic regions of specialization. Colonial science, if studied in semi-autonomous disciplinary formations that fail to engage with the central claims of Western history and sociology of science, will simply continue to insulate European histories of science from the need to consider the ways in which non-Western knowledges are constitutive of notions considered native to English, French, or Dutch science.

Although most histories of colonial science demonstrate primarily its effect on the disciplining of colonial subjects within one region, they contain elements of a history of global networks which suggest possibilities for enriching our historical understandings of globalization and cross-cultural exchange (processes commonly spoken of today as if they were novel creations of the telecommunications revolution). Histories of British imperial botany demonstrate that botanical scientific theories as well as people—scientific administrators, agricultural researchers and extension workers, botanical explorers, gardeners, and plantation laborers—were global travelers, moving among the numerous colonial botanical stations, or between Kew and scientifically uncharted regions outside the British Empire. The collection of botanical specimens created human and political networks across geographic boundaries within and among different European global formations. Early environmentalist theories, based on agricultural and forestry research, were developed through intellectual exchanges among British, French, Dutch, and German scientists and administrators. The story of the civilizing mission in India is not tied exclusively to imperial Britain, either: it included not only Anglo-Saxon Protestant missions, but German Protestant enterprises and Catholic missionaries from all over Europe. Then, as now, questions of science, economics, power, and culture were worked out in the context of the cross-border rhetorical and material practices that constituted the politics of globalization. Sustained engagement between metropolitan and colonial histories can offer more than a casual interdisciplinary encounter; it can reveal wider connections among global actors in the past, and yield richer analytical perspectives on the present phase of globalized exchanges.

Eric Wolf, in *Europe and the People without History,* radically critiqued the division between histories of Europe and accounts of other cultures, arguing for the writing of global histories which acknowledge that the two have long been part of the same manifold.[17] The suppression and omission of non-Western histories that Wolf noted in his 1982 book was rectified to a large extent in the decades that followed; nevertheless, a kindly cultural relativism often results in the relegation of "other" histories of knowledge to the realm of ethnosciences and histories of the margin. Some theorists in science studies and non-Western history celebrate the margins by way of a postmodernist relativism, valuing the local over the global in a belief that this will subvert "metanarratives" such as those of universalist scientific objectivity. Others, however, argue that such a relativist position encourages an abstract, removed, separate-but-equal kind of tolerance rather than a noncolonizing dialogue between cultures.[18] Philosophers of science such as Richard Boyd and Satya Mohanty argue against epistemological relativism and in favor of a "post-positivist realism" in science studies and postcolonial studies. Such a position is, they argue, based on the assumption of a common ground between self and other, and a continuity rather than a radical difference between Western and indigenous knowledges.

A Note on Context and Method

The discourses and practices analyzed in this book—state-run scientific forestry, entrepreneurial planting, technologically utopian Christianity, and the science of anthropology—were linked by material conditions and discursive patterns that make it important for us to analyze them not as autonomous spheres but as elements of a single process.

The methods of cultural materialist historiography implicitly counter idealist critiques of modernity by showing that colonial scientific modernity was made up of a mixture of contradictory elements of indigenous labor, resistance and cooptation, religiously inflected moral agendas, and an Enlightenment science that was often seen as a protector, and not merely an exploiter, of nature. The contradictions among these elements can be explained in terms of their functionality within an emerging global capitalist system of resource extraction, production, and distribution. If we are to understand the functioning of modernity in postcolonial societies today, the supposedly premodern elements must be acknowledged and theorized as functional elements of this modernity rather than regarded as aberrations or vestiges of an old system that will wither away by itself.

Implicit in this model of mixed modernity is the methodological injunction to pay attention to the varied responses of different sections of colonial society to scientific modernization. Indian colonial society was not homogeneous—an assumption we risk making if we focus only on the expression of colonial policies or prejudices. There were preexisting stratifications along caste, region, community, and gender lines, which were altered or made more rigid during the nineteenth century. Tribals and Brahmins encountered different faces of colonial scientific modernity, and had radically different modes of response open to them. There remains a need for detailed studies of the different aspects of modernity and responses to it that pay attention to differences along gender, caste, and regional divisions.

By interrogating a historical account of science with questions explicitly formulated in the context of current theoretical and political concerns, we can lay the groundwork for a better understanding of postcolonial scientific modernity. To carry out nonreductive analyses of postcolonial scientific discourses of collaboration and resistance will require us to transcend traditional disciplinary boundaries between history, science studies, and cultural studies so as to formulate more adequate alternatives to conventional narratives of progress. I hope, by transgressing some of these disciplinary boundaries here, to support the new directions that are beginning to open up for the analysis of science and society in South Asia.

There is, of course, a tension between "knowing the past in its own terms" and the anachronism of inserting "present concerns into the study of the past."[19] I find Dominick La Capra's characterization of interpretive history as a "dialogue with the past" useful in explaining the tensions and rewards in a historiography that owns up to both contemporary theoretical interests and a concern with historical accuracy. He argues, in *Rethinking Intellectual History*, that "what underlies the concept of anachronism . . . is the idea that a radical divide separates periods and times and that the only alternative to an abstract concept of difference over time is an equally abstract concept of intemporal universality (or human nature). . . . [T]he concept of anachronism, whatever its limited usefulness, must not itself become a bar to a dialogue across time that is itself respectful of differences and distances—a dialogue that even those relying greatly on the concept of anachronism acknowledge in practice."[20] La Capra argues that we need to develop theoretically informed modes of historical investigation that combine the best methods of social history and textual interpretation. The metaphor of dialogue with the past serves to emphasize the interpretive stance of the historian toward her sources, which might be texts or artifacts with multiply layered, hidden, or marginalized meanings.

It is in this sense that I offer close readings of scientists' accounts and popular discourse, in conjunction with an attention to social practices in their larger political context. The history of colonialism, the process of decolonization, and the condition of postcoloniality have been widely written about in the last two decades.[21] A cultural materialist study of colonial science can and should transcend traditional boundaries between environmental studies, cultural studies, and colonial historiography, to engage in a dialogue with the past—one that combines political engagement and cultural critique without abandoning the rigor of complex and detailed historical analysis.

Postcolonial Historiography and the Pitfalls of Dialogue with the Past

> Tribes which, only a few years ago, were living in a wild state, clad in a cool and simple garb of forest leaves, and buried away in the depths of the jungle, have now come under the domesticating and sometimes detrimental influence of contact with Europeans, with a resulting modification of their conditions of life, morality and even language. The Paniyans of the Wynad, . . . the Irulas who inhabit the slopes of the Nilgiris, now work regularly for daily wage on planters estates; and I was lately shocked by seeing a Toda boy studying for the third standard, in Tamil, instead of tending the buffaloes of his mand. The Todas, whose natural drink is milk, now delight in bottled beer, and mixture of port wine and gin, which they purchase in the Ootacamond bazaar.[22]

When Edgar Thurston wrote these lines in the last years of the nineteenth century, he was one of the most eminent among the British ethnographers who spent their lives studying Indian tribes and castes. He recounted with disapproval the changes that turn-of-the-century tribal India was caught up in.

> A Toda lassie curling her ringlets with the assistance of a cheap looking-glass; a Toda man smeared with Hindu sect marks, doing puja, and praying for male offspring at a Hindu shrine; a Bengali babu with close-cropped hair and bare head, clad in patent leather boots,

> white socks, dhoti, and conspicuous unstarched shirt of English device; a Hindu or Parsi cricket eleven engaged against a European team; the increasing struggle for small-paid appointments under Government: these are a few examples of changes resulting from the refinements of modern civilization.[23]

The romanticist critique of modernity affects a nostalgia for a prescientific past, and sees the evils of modernity in the hybrid appropriations of modernity by natives. This hybridity is seen as absurd and dismaying (sometimes threatening) because the natives are somehow, the romanticist believes, inherently nonmodern, hence can never quite fit with ease into modern ways (hence the inevitably inappropriate white socks and crushed shirt, the ignoble struggle for petty jobs, or the startling juxtaposition of heathen religions and cricket whites).[24] The "premodern savage" is valorized in opposition to the "English-educated native clerk."

> The Kadirs afford a typical example of happiness without culture. Unspoiled by education, the advancing wave of which has not yet engulfed them, they still retain many of their simple manners and customs. Quite refreshing was it to hear the hearty shrieks of laughter of the nude curly-haired children, wholly illiterate, and happy in their ignorance. . . . The uncultured Kadir, living a hardy outdoor life, and capable of appreciating to the full the enjoyment of an "apathetic rest" as perfect bliss, has, I am convinced, in many ways, the advantage over the poor underfed student with a small-paid appointment under the Govt as the narrow goal to which the laborious passing of examination tests leads.[25]

The nostalgic critique of modernity obscures both the instrumental rationality of indigenous epistemologies (as it effectively relegates native knowledge of nature to nonrational cosmologies), and the political and economic basis of modernity (as it represents ideal modern life as a set of practices that accord with notions of high culture; the native appropriations of these are thus seen as low-class aspirations to sophistication). Modern-day romanticist appropriations of indigeneity easily fall into a similar dichotomy, as they invariably fail to apprehend the material conditions that make one epistemology unviable, and another take precedence.

In boardrooms, classrooms, and nationalist festivals, the call to celebrate, commercialize, preserve in museums or otherwise render static the antimodern Other resounds forcefully today. The idealist desire for a lost past and its accompanying nostalgic critique of scientific modernity has many resonances with the discourse of nineteenth-century romanticist ethnography. Any romantic critic today who believes herself to be defending pure indigenous pasts against destructive Western modernity need only look back about a century or so to find soul mates among colonial anthropologists who mourned the loss of primitive cultures. In contrast to a supposedly inherently destructive Western science and a monolithic Enlightenment rationality, indigenous people are construed as inherently ecologically wise, and as having always lived in harmony with nature.[26] Critics of modernity imagine industrial society's vices to be transcended by an indigenous essence, the appropriation of which may deliver us back to a state of oneness with nature. They often argue that the rosy conditions of the past can be regained if we strip off the overlay of scientific modernity that has been imposed on non-Western societies. This argument often fails, however, to take account of two domains: the complex internal stratifications and conflicts within the non-Western society under consideration, and the mixed nature of scientific modernity itself, involving the incorporation of supposedly premodern elements into what we consider "modern" formations.

The following chapters will take up a range of examples of the construction of modernity in nineteenth-century British India. Examining the social and political basis for diverse official and lay understandings of scientific progress may help demystify the process of modernization, which has continued beyond independence with many colonial dichotomies intact or simply inverted but not dismantled. These dichotomies are part of systems of knowledge that have been naturalized, not by sweeping away what came before but by becoming imbricated with it.

A materialist understanding of scientific epistemologies would suggest the probability that the indigenous knowledges displaced by scientific modernity were themselves perfectly rational and efficacious within a sociopolitical system (or mode of production with its accompanying social relations) that was suddenly rendered obsolete.[27] If a reconstruction of such knowledges is to take place, it must be done with a full analysis of their now-compromised socioeconomic conditions of possibility. The careful analysis of the cultural and political underpinnings of local knowledges is not equivalent to an uncritical valorization of "authentic" or "purely indigenous" ways of knowing. It can, however, work toward redressing the ways in which both indigenous and Western epistemologies were (and continue to be) evaluated, by deconstructing the hierarchy whereby the former are regarded as inherently unscientific as a result of being embedded in sociality, and the latter naturally superior as a result of transcending social and political interests.

Reconstructions of local knowledges can serve as a useful exploration of alternatives to the present rapidly changing conditions of modernity. The rhetoric of scientific advance is even today couched in terms of a naturalized progression from superstition to free markets. A recognition of the historical interaction between social and environmental change will influence our political and economic prognosis of the future, at the very least making us realize that it is impossible for us either to go back or to go forward without incorporating a model of the social, political-economic, and ecological relations that have been transformed. A historicized ecology makes an idealized nostalgia difficult to sustain, as it becomes clear that no linear reversal of steps is possible in order to regain a pristine origin. Such studies can work in anti-idealist solidarity with present-day efforts to find a political voice for underrepresented knowledges, by pointing out the material and ideological processes one must take into account in order to see systems of knowledge about nature as neither mystically rooted in tradition nor serenely transcendent of political interests.

Nature's Colonial History

"He lived like the wild animals he hunted, and was almost as wild. . . . I was the first white man he had seen. . . . A dose of rum and a few rupees made him most amenable," wrote Brigadier-General Reginald Burton in 1928.[28] Notions of wildness both as resource and as threat to civilization undergirded European philosophies of science, technology, and modernity. British colonial writings are replete with descriptions of wild landscapes and peoples, unwitting windows into the ideological frameworks of their time. Intriguing patterns of history emerge if we trace the intersecting loci of wild and tamed, enraged and docile, slothful and disciplined, productive and primitive, superstitious and scientific nature and culture.

Reminiscing about his hunting activities in late-nineteenth-century India, Brigadier-General Burton recalled his "tireless" and "trustworthy" local guides, such as the "wild Gond shikari" described above. Hunting helped construct both ecological and social norms in many colonial contexts. Many missionaries to Africa financed their expeditions with ivory and offered African populations supplies of meat as incentives for labor.[29] In the second half of the nineteenth century, hunting in colonial Africa became ritualized and consecrated as The Hunt.[30] While they decimated wildlife in Africa, white hunters

FIGURE 1. Lord and Lady Curzon at a tiger hunt during a visit to the Nizam of Hyderabad, 1902. From Ray Desmond, *Victorian India in Focus: A Selection of Early Photographs from the Collection in the India Office Library and Records* (London: Her Majesty's Stationery Office, 1982), frontispiece. *By permission of the British Library.*

became renowned as conservationists, natural historians, and role models for boy scouts in England. John MacKenzie cites Jason Machiwanyika, an African in early twentieth-century Rhodesia: "Europeans took all guns from Africans and refused to let them shoot game. But Europeans shoot game. Africans have to eat relish with only vegetables. If an African shoots an animal with a gun, the African is arrested and the gun is confiscated."[31]

In periods of famine, Africans no longer had recourse to game meat as a source of protein, because of both their lack of weapons and the scarcity of animals. Although white hunters were contemptuous of the hunting abilities (and hence the "manliness") of most Africans, they valued "Bushmen" as guides and trackers and depended on their expertise. Several colonial campaigners compared suppressing Africans with hunting animals. For example, E.A.H. Alderson, in his 1900 book on the Shona uprising, compared shooting the Shona to "chasing the fox, rabbiting from bolt holes, shooting snipe, and scaring rooks."[32]

In India, although the British use of tribals as trackers recalls the similar use of Bushmen in Africa, there was an elite dimension to the social significance of hunting, *shikar.* British hunters were often hosted by the rulers of princely states, where hunting was a traditional royal pastime, and there were few game regulations that applied to such hunting parties. Here feudal and colonial rituals reinforced each other, and together had consequences for the survival of wildlife and for peasant and tribal access to game.

Forester and historian E. P. Stebbing, like many other colonial officials, wrote a book of hunting memoirs, in which he declared: "the first golden rule for the sportsman to bear in mind is that he will get far more sport and a much greater enjoyment out of it if he has the villagers amongst whom he is living on his side."[33] Indigenous groups were represented as having accurate knowledge of nature due to the wild or natural state in which they lived. Whereas representations of royal shikar always portrayed native princes as exotic, powerful dominators of the forests and game, representations of tribal shikaris were rendered in terms that explicitly equated tribal with animal existence. Natives themselves were repeatedly constructed as part of nature. When we read nineteenth-century accounts of nature we are invariably also reading colonial representations of natives.

Such a construction of locals, drawing on a nature-culture dichotomy and its attendant hierarchies, allowed colonial officials and scientists to perceive themselves as inherently superior to indigenous tribal inhabitants of colonial societies. Such constructions also enabled colonial officials to contain the threat of the savage, hostile native by representing the native (like nature) as useful if well managed, and as dangerous or useless if uncontrolled.[34] Nor was this only a rhetorical parallel. As I show in chapters 3 (on forestry), 4 (on plantations), and 5 (on anthropology), these constructions were enabled by material conditions and had practical consequences for local lives.

In nineteenth-century India, as in colonial Africa, hunting and conservation were often advocated by the same groups of people. Foresters and planters in India were generally avid hunters, and belonged to hunt clubs that had interests in conserving game so as to have a constant supply for their members. At the same time, foresters and planters were charting out elaborate systems of land use that were legitimated by commercial as well as climatological arguments.[35]

There was considerable intellectual exchange between scientific institutions in London (such as Kew Gardens and the British Association for the Advancement of Science) and doctors, geographers, foresters, and planters in the colonies. In the Bombay Natural History Society, founded in 1883, the scientific interest in fauna was combined with the hunting lobby for game

HOT WEATHER—THE MIDDAY HALT

RIFLE AND
ROMANCE
IN THE INDIAN JUNGLE
A RECORD OF THIRTEEN YEARS
BY CAPTAIN A. I. R. GLASFURD
OF THE INDIAN ARMY WITH
NUMEROUS ILLUSTRATIONS BY THE
AUTHOR AND FROM PHOTOGRAPHS

LONDON: JOHN LANE THE BODLEY HEAD
NEW YORK: JOHN LANE COMPANY
TORONTO: BELL & COCKBURN. MCMXIII

FIGURE 2. Sample of a popular genre: the hunting memoir. Title page and frontispiece, Captain A.I.R. Glasfurd, *Rifle and Romance in the Indian Jungle* (London: John Lane the Bodley Head, 1913).

conservation.[36] The south Indian planters, too, formed a game association that protected wildlife habitats to preserve game species for their sport. Whereas royal shikar had been the most common form of the sport in the early nineteenth century, by the end of the century British officials were no longer dependent on Indian princes and kings for access to rich hunting grounds, as they controlled large portions of the forested regions of the subcontinent.

State control of forest and water resources ultimately led to a rationalization of land use in accordance with the changes in the nineteenth-century global political economy. As Richard Grove has pointed out, colonial scientists and administrators initiated a discourse of conservation that took into account the need to preserve natural resources in a rapidly changing, globally networked colonial economy.[37] In this discourse, however, the commercial and imperial motives for conservation often took precedence over the customary rights of local users. Forest laws in Indian and African colonies invariably dispossessed tribals and subsistence peasants of their rights to the land. Local methods of resource use (such as slash-and-burn agriculture, the use of village commons for grazing, hunting, and the gathering of forest products for subsistence) were subordinated to metropolitan demands on the land. Native populations were organized into labor pools to service this metropolitan demand. Sedentarized tribal populations were studied by ethnographers, managed by foresters and revenue officials, recruited by plantation labor contractors, and preached to by missionaries. The links between foresters, geographers, planters, missionaries, and ethnographers reveal intersecting discourses that sustained colonial natural- and human-resource policies.

Nineteenth-century tropical forestry, natural history, and ethnography were scientific disciplines that were to different degrees dependent on the political context of colonialism. The importance of each of these three sciences to the state can be demonstrated by an examination of the administration reports of the colonial state for any of the years of Crown rule.[38] They each include, among the numerous administrative categories, a section on botany or natural history (the terms vary with the period), an extensive report on forestry, and a section on aboriginal tribes that dealt with ethnographic knowledge of specific tribes and reported incidents of crime, discussing consequent policy directions for law and order in the Agency Tracts (forested regions populated almost exclusively by tribals). These were linked with each other rhetorically and ideologically.

Natural history after Linnaeus had relied heavily on the accounts of European travelers in Asia for the compilation of comprehensive lists of species. When exploration and trade were succeeded by direct colonial rule in the eighteenth and nineteenth centuries, scientific interest in Asia grew; colonialism altered scientific spheres of interest and methods of observation and systematization. Nineteenth-century colonial concerns about the stability of the empire, and interest in the commercial application of science and technology within the expanding global market, supported the creation of scientific knowledge that would benefit the state in both these contexts. The gathering and ordering of scientific facts became an enterprise systematically fostered by the state. The representations of nature, animals, and people this process engendered tell the historian as much about the politics of colonialism as about local ecologies. By examining botanists' and foresters' descriptions of nature, we can see how a narrative of progress was constructed wherein the scientific and conservationist colonial administrator brought order to a chaotic and threatening tropical landscape peopled by ecologically profligate natives.

There were numerous episodes of everyday resistance to state control of nature. This resistance took the form of forest theft, arson, and the evasion of surveillance throughout the forest tracts of Madras Presidency. These did not, however, erupt into full-scale rebellions, as they often did in neighboring presidencies. This was partly due to the existence of several "private" or unreserved forests from which forest groups could fulfill their basic needs, and to the Madras government's attempt to serve customary forest rights; but it was also due to the forest department's successful surveillance of the native population and the systematic incorporation of tribals and landless peasants into a system of wage labor that demanded the sedentarization and disciplining of mobile nomadic populations. The Criminal Tribes Act of 1871 was a crucial piece of legislation in this process, and relied for its information on ethnographic data about the criminal proclivities of specific tribes. Under this act, entire tribes could be classified as essentially criminal. Such a generalization was implicitly undergirded by the epistemological framework offered by colonial anthropology, which made it possible for colonizers to claim knowledge of inherent characteristics of particular (racialized) groups of natives. The legal classification allowed police and administrative authorities to restrict severely the movements of individual members of the tribes.[39] The forced sedentarization of shifting cultivators and nomadic populations facilitated more efficient surveillance of these tribes, some of whom, alienated from their traditional hunting privileges or social roles, often resorted to petty crime as a means of survival. Official literature on this resettlement process emphasized the importance of introducing "primitive" and "undisciplined" tribals to the ethic of wage labor, private property, and stable residence. As part of the program, tribals were required to take up agricultural activities or gather forest produce as employees of the forest department.

In this process, then, it is possible to see the ideological aspect of three practices functioning simultaneously and in symbiotic relationship with each other: the anthropological scheme of scientific hierarchies of mankind, the commercial aspects of agricultural and forest science, and the legal machinery of the state.

Natural history, anthropology, and forestry were intimately linked not only with each other but with global political economy and colonial state policy. The connections can be seen most clearly by examining the rhetorical constructions of nature and of the native as they became objects of scientific study and commercial management, rhetorics that display curiously similar patterns. Commercially motivated enterprises such as the cultivation of cinchona and the establishment of rubber, coffee, and tea plantations drew their legitimation not only from a contemporary environmentalist discourse but also from a "scientific" model of society's progression from the hunter-gathering, through shifting cultivation and settled agriculture, to the settled urban stage. The elaboration of these intersecting processes is one of the threads that links the following chapters.

Labor recruitment for the plantations and the disciplining of the worker were legitimated on an allegedly scientific hierarchy of societies and economic structures. A reading of the missionary archives and plantation records indicates that missionaries were encouraged to bring their civilizing message to plantation workers' homes on the premises of privately owned tea and rubber estates.[40] The plantation economy had wide-reaching environmental and political effects—the former connected with the replacement of tropical forest with monocultures, and the latter connected with the creation of a new class of landowners whose economic power in the region continues, only slightly abated, to this day.[41]

From a cursory perspective, these five categories—natural history, ethnography, forestry, plantations, and Christian missions—might seem unconnected. Showing that they were linked by diverse discourses and practices, however, and discerning theoretical and ideological continuities among them, I suggest that we can unite the theoretical insights of postcolonial historiography and of science studies, in order to elucidate how ideologies of modernity functioned both at the level of state policy and at the level of everyday experience.

This book, then, examines the practices and theory of two kinds of "scientific" enterprises (ethnography and forestry) and two "nonscientific" enterprises (private plantations and Christian missions) in the context of colonial south India.[42] These enterprises gained legitimacy not only through their material effects but also through drawing upon a constructed notion of the scientific. Assuming that science is a form of culture that is continuous with rather than radically distinct from other social practices, chapters 4 and 6 trace a scientific rhetoric of progress and modernity in the nonscientific practices of entrepreneurship and proselytization. Taken together with the culturalist rhetoric in discourses of scientific forestry (chapter 3) and the science of man (chapter 5), they illustrate how the boundaries between scientific and nonscientific discourses are negotiable, and how, in the politically charged context of colonialism, the rhetorical and material traffic between science, culture, and commerce had significant implications for both the social construction of modernity and the institutional practice of science. Interdependent constructions of nature, natives, and scientific modernity link the disparate discourses of chapters 2 through 7. Chapters 2 and 7 bookend the explorations of individual discourses in chapters 3, 4, 5, and 6, by offering two stories in which multiple discourses of nature, culture, science, and progress are jumbled together. Rather than disentangling these discourses, these two chapters highlight their interdependence. Chapter 2 is a local story that focuses on these entangled discourses and practices in a small hill town and its surroundings, whereas chapter 7 is a global story focusing on these entanglements within global networks of commerce and knowledge.

A Note on Indian History

Since each chapter of this book follows discursive themes, analyzing different processes between 1858 and 1930 rather than following a strict chronological structure, the analysis might be confusing to the reader unfamiliar with standard histories of colonial India. This section offers, therefore, a thumbnail sketch of colonial history in India.

Although the content of the following chapters focuses on the late nineteenth and early twentieth centuries, it is worth bearing in mind that British colonizers did not encounter a homogeneous subcontinent untouched by global influence. The period prior to the nineteenth century remains largely unexplored by postcolonialist cultural critics, although a large body of economic and political scholarship exists on earlier periods.[43]

Southwest India participated in global trade and culture for several centuries before the advent of British rule in India. Even before Vasco da Gama landed at Calicut in 1498, this port on India's southwest coast was one of the major entrepôts for the extensive Indian Ocean spice trade, which linked India's pepper and spice production to the ports of Malacca, Hormuz, and Aden, across Egypt to the markets of Alexandria, and thence to Venice and other European markets. The Indian Ocean trade created a vast multiethnic economic formation, which was largely open and tolerant in terms of trade practices, although pirates and merchant groups certainly exerted their power.[44] The system was largely free from specialized political control until it was modified first by the Portuguese and then by the Dutch, both of whom systematically attempted to expel other traders (especially Arab Muslims) from the area. M. N. Pearson records that the ruler of Calicut refused an early Portuguese attempt in 1502 to expel the Muslims, finding it "unthinkable that he expel 4,000 households of them, who lived in Calicut as natives, not foreigners, and who had contributed great profits to his Kingdom."[45] The Portuguese did achieve economic dominance over the global spice trade, but never quite wiped out local initiative, as indicated by the surreptitious trade that traders in Malabar and Kanara carried on successfully during the sixteenth century. Portuguese power was eclipsed by the Dutch East India Company in the early seventeenth century, which itself went bankrupt in the 1790s, but only after having controlled a vast South and Southeast Asian trade in spices and slaves.[46] Meanwhile, in the first half of the eighteenth century, the British and French consolidated their trading companies in India, clashing several times in mid-century in their global tussle for supremacy.

Southern India, on which the following chapters will focus, had been accustomed for centuries to economic exchange and cultural transmission, and included cosmopolitan and mobile trading populations, including Malabar traders with links to North Africa and the Middle East, and Chettiar traders who had links to Southeast Asia. The lasting effects of colonial rule on Indian developmental policy stems not from a disruption of a supposed precolonial cultural purity nor from the subcontinent's mere insertion into a global network of trade controlled by a new world power. Rather, scientific modernity began to be thought about and practiced in new ways, shaping individual and national attitudes toward the home, the world, and the relationship between them. The emergence of post-1857 Indian modernity transformed the ways both Indians and Englishmen conceived of the self, the Other, and civilizational progress. Inevitably implicated in these shifts were the patterns of the colonial state's administrative agenda.

In hindsight, British rule in India is usually dated from 1757, when Lietenant-Colonel Robert Clive defeated Siraj-ud-Daula, the nawab of Bengal, by arranging the defection of the nawab's commander of troops at the

Battle of Plassey. At the time this appeared to the reigning Mughals to be a regional skirmish, resulting in the not-unwelcome defeat of the almost-independent nawab. They offered the civil administration of Bengal to the British. Although Clive wanted the British Crown to accept a responsibility he saw as too great for the British East India Company, Prime Minister William Pitt rejected the idea, fearing that so much wealth in the hands of King George III would eclipse the power of Parliament.[47] Clive became governor of Bengal, beginning a century of "Company rule" that enriched numerous adventurous Englishmen beyond their wildest dreams. Trained at Haileybury College in Malthusian theories of population and Ricardo's theory of rent, British administrators refined land revenue systems so as to extract agrarian revenues far more efficiently than the Mughals, who had been accustomed to accommodate climatic variation and other contingencies that altered their incomes from year to year.

The East India Company's monopoly on Indian trade was abolished in 1813, after which many other private agencies shared in the profits. In 1833, Parliament dissolved the East India Company as a trading house, but the Company continued to work like an export bank, and continued to finance the transfer of money from India to England.[48] In the 1850s the railways became the most spectacular technological mode of managing the subcontinent. Company rule ended with the 1857 Mutiny (or First War of Independence), after which the Crown officially took over Indian rule. The collaborationist English-educated elite, who had remained aloof from this peasant rebellion that almost ousted the British, were rewarded by the new conservatism that characterized Crown rule under Queen Victoria. Believing that their liberal experiments in rule had failed to win over an ungrateful populace, Victorian Britain began, through an intervening class of educated indigenous elites, to exert a more coercive order on its colonial subjects, while intensifying the extraction of raw materials for its own growing industrial economy.

The early nineteenth century had also seen fierce competition among rival British political philosophers and economists, who saw India as an administrative laboratory for the vindication of their theories. The controversy between Anglicist and Utilitarian versus Orientalist theories of education, for example, is best known through Thomas Babbington Macaulay's 1835 Minute on Education, in which he deemed all of Indian scholarship worth less than a single library shelf of Western knowledge. As Penelope Carson has noted, many historians who cite this document to indict the Anglicists forget to note that Anglicists and Orientalists in fact agreed on the putative superiority of Western science and literature. The disagreement was over whether vernacular languages and literature or English education was the most efficient route to educating Indians in such a way that they might facilitate the colonial government's task of daily administration.[49]

The English-educated elite began to form nationalist associations in the 1870s, holding the first annual session of the Indian National Congress in Bombay in 1885. Liberal nationalists sought social reform within the context of continued British rule in India. Although the partition of Bengal in 1905 spurred a more revolutionary brand of religious nationalism (such as that inspired by Swami Vivekananda), which sought spiritual and political separation from the British, liberal nationalism remained more influential among south Indians. Eric Stokes accounts for liberal nationalism's ultimate success by pointing out how little it challenged the British conception of governance.

> From the mid nineteenth century this [Western-educated] element came to man the middle and upper levels of the increasingly technical, legalistic, and bureaucratic machine that was the Raj; and from the 1880s slowly began to secure effective command of the channels of communication between the Raj and the millions of its subjects. The British thus became steadily more dependent on the modernistic section among their collaborators. In consequence, what is usually thought of as the history of the nationalist struggle was little more than the struggle of this section to extort from the British the loaves and fishes of political office and administrative place. . . . The ultimate victory of Congress should not be allowed to disguise the fact that what it represented was, on this view, not the victory of a new class but the lateral adjustment of an old elite.[50]

Some Indian Marxist historiography has proffered similar views, stressing the comprador function of liberal nationalists, and noting that colonial rule enjoyed a symbiotic relation with feudalism.[51] In an important contribution to the history of science, Gyan Prakash in *Another Reason* has noted the contradictory nature of Indian modernity in terms of the liberal nationalists' enthusiastic participation in technological development and the accompanying, ever-more efficient and exploitative, modes of extraction and production.[52]

It is clear that modernity had contradictory effects on the urban and the rural, on the home and the outside world, and on state policy and popular culture. Since nationalism embodied these contradictions, it is unsurprising that many of these contradictions persisted beyond 1947, when the proverbial sun finally set on the British Empire in India. In conducting an investigation of science in colonial India, then, it is appropriate to adopt an anthropological attention to local detail, material and cultural constraints, and everyday practices, while keeping in mind the larger picture of global trade, colonial politics, and nationalist ideology. This is a larger canvas than is usually admitted into a work of science studies, but only, perhaps, because Euro-American science studies can assume the national historical contexts of British or American science to be well known. In extending science studies to non-Western contexts, we are faced with the opportunity and challenge of synthesizing the work of several disciplines, while retaining an investigative focus on the operation of science as idea and practice.

CHAPTER 2

A Local Story

ENGLISH MUD

When we ask questions about the mutual constitution of colonial science and culture in colonial India, we encounter myriad discourses and practices. Focusing on the Nilgiri Hills of southwestern India, this chapter offers an overview of the intersecting colonial narratives of natural and cultural progress by describing the dramatic changes in the physical and conceptual landscape of these hills after they were discovered by the British. Examining the discourses of English administrators, ethnographers, missionaries, planters, foresters, and hunters, we can read their intersecting narratives of nature in order to understand not only attitudes toward nature in a particular historical context but also the ways of knowing they legitimate and the complex power relations within which these different narratives intersect, complement, or conflict with one another.

Lord Lytton wrote to his wife while staying at the Government House at Ootacamund (known popularly as Ooty), in the Nilgiri mountains of the Western Ghats, describing the summer capital of Madras: "I affirm it to be a paradise. The afternoon was rainy and the road muddy but such beautiful *English* rain, such delicious *English* mud!"[1] He went on to describe Ooty as a combination of "Hertfordshire lanes, Devonshire towns, Westmoreland lakes, Scottish trout streams, and Lusitanian views."[2] Thomas Macaulay, after a visit to Ooty in June 1834, wrote that it had "very much the look of a rising English watering place." He reported to his sister that the scenery looked like "the vegetation of Windsor forest or Blanheim spread over the mountains of Cumberland."[3]

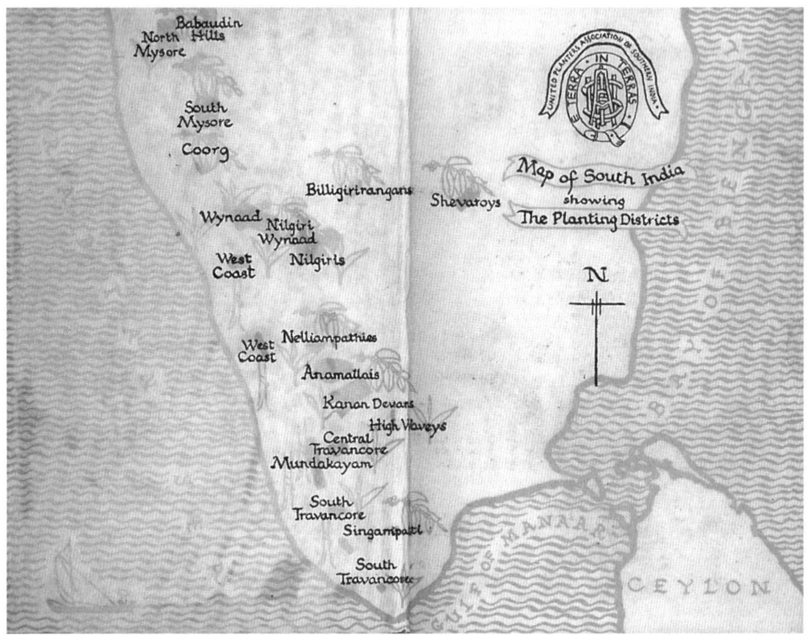

FIGURE 3. Map of South India. From S. G. Speer, *UPASI: 1893–1953*. Diamond Jubilee Publication (Coonoor: United Planters' Association of Southern India, 1953).

Lytton and Macaulay were only two of the most eminent of the hundreds of English residents of India who in letters, magazines, guidebooks, and histories extolled the English virtues of Ootacamund.[4] Was this narrative of the Englishness of the landscape a consequence of the "naturally English" landscape of the Nilgiri Hills? I will argue, rather, that it was a culturally and politically significant construction of the environment, and that in examining such narratives of the environment, we can arrive at a method by which to study the intersecting mechanisms that create the politics of nature. The following parallel narratives of the Nilgiri Hills are instrumentally rational and internally contradictory; and these inconsistencies tell us something about the ideological basis of English presence in the hills.

The Nilgiri District of Madras Presidency, where the British colonial government located its summer capital in order to escape the heat of the plains, was the object not of those common constructions of the colonized landscape as hostile and barren, so familiar to Western audiences through the films *Passage to India* or *Heat and Dust,* but of constructions so cloying and rosy that they might seem, on a first reading, to assimilate India to the English self, rather than figuring it as radically other. The first section of the chapter will interrogate these constructions, moving from representations of nature in descriptions of climate, vegetation, landscape, and so on, to representations of power and authority in the management of nature and from there to the interwoven management narratives of nature and natives. The second section takes up the possibility of counternarratives. Any investigation of hegemonic constructions cannot help but reveal the ways in which these are rendered unstable even in the act of construction. This short section will identify some of the unstable moments in colonial constructions of the Nilgiri inhabitants, suggesting where the fractures lie in what might otherwise come across as a seamless and closed narrative.[5] The third section, finally, surveys the diverse array of narratives that come under our scrutiny as we refine and expand our analytical lens.

Ootacamund: English Eden in the Nilgiris

The first record of European interest in the Nilgiris is in 1602, when Francisco Roy, the first Roman Catholic bishop of the Syrian Christians, dispatched a priest and a deacon to investigate stories of a lost tribe of Christians "in a country called Todamala," with the aim "that priests and preachers should be sent thither immediately to redeem them to the Catholic faith, baptise them, etc." They found the Todas, whom they believed to be degenerated from "the ancient Christians of St. Thomas." The priest reported that he conversed with a Toda priest and presented "looking glasses and hanks of thread" to Toda women, but returned with "no information leading to the supposition that either they or their ancestors ever had anything to do with any form or profession of Christianity."[6] There was almost no European interest in the region for the next two hundred years. From 1799 onward, the Nilgiri mountains were included in the revenue area of Coimbatore, and Price reports that a Chetty would periodically collect revenue. In 1819, two Europeans in pursuit of a Poligar accused of robbing Coimbatore peasants followed him through a mountain pass and discovered what was later to become the town of Ootacamund, then a Toda settlement.[7]

John Sullivan, collector of Coimbatore, was very excited by the find, and in two years had built his own residence there. Here he spent increasing amounts of time, reluctant to leave a land which he envisioned as a home away from Home, a promised land where Europeans would multiply and thrive on English produce. One of the first things he did was to import a gardener from England, to make this vision into a reality. One commentator records:

FIGURE 4. Members of the Toda tribe. From Ray Desmond, *Victorian India in Focus: A Selection of Early Photographs from the Collection in the India Office Library and Records* (London: Her Majesty's Stationery Office, 1982), 54. *By permission of the British Library.*

> [G]ardening . . . was a passion with Mr. Sullivan. He sent for a gardener from England, a Mr. Johnson, who was left in charge of building operations when the Collector had to tear himself reluctantly away to attend to the affairs of his district. With Johnson, in 1821, arrived the first English apple and peach trees and strawberries, the first seeds of flowers and vegetables. . . . Potatoes were introduced, and flourished. . . . A few years after Mr. Sullivan's arrival, gargantuan wonders were being compared as proudly as though their owners were competing in a village show at home. A beet is recorded as being nearly three feet round, a radish three feet long, and a cabbage plant eight feet high. Geraniums grew in hedges, and somebody's verbena forgot its place and shot up to the sky on a trunk like that of a robust tree. English oaks and firs were planted . . . in 1829 there were wild white strawberries, wild Ayrshire roses, and small,

> deep damask roses growing in Ooty. . . . A convalescent young officer . . . [described Ooty as] "presenting to the eye a wildered paradise."[8]

The specially imported Johnson and a host of enthusiastic fellow gardeners rapidly transformed Ooty into an English vision of paradise, refracted through colonialist tropes of gigantic tropical wonderlands, thus ending up with an image of Home enlarged into fantastic proportions.[9]

Ooty, like other English settlements in the hills, grew around state government official activities and the stationing of troops in a climate conducive to English patterns of life. As Anthony King has noted, the hill station "provided, in its physical, social, psychological and 'aesthetic' climate, the closest approximation to conditions of life 'at Home.' Of all areas of settlement, it was the hill station which most easily permitted the reproduction of metropolitan institutions, activity patterns and environments."[10] Although the attraction of the hills came initially from what seemed to be their astonishing natural climatic and topographic similarity to England, King goes on to note, they came increasingly to be cherished as sites of the elimination of two factors that "restricted social life" in the plains. First, because of their relative isolation, the hills could not be monitored as easily by British officials, whether in India or in London. Thus an individual English officer's authority was practically unlimited in the hills. Second, English officers and their families could retreat from the large urban Indian populations of the plains to the sparsely populated hills, in which natives could be seen as quaint rather than threatening.[11]

The overcoming of three major constraints of the plains (oppressive climate, restricted authority, and ubiquitous natives) is the basis of much narrative building by Europeans in the Nilgiris from the founding of Ootacamund in 1821 up through the twentieth century (and, in fact, through to the present). The following sections introduce these narratives.

Climate and Nature

Consider the following early twentieth-century verse entitled "Nilgiri Sunshine," penned by a civil service officer in Ooty.

> Outside the Clubhouse
> The old fellows sit
> Feeling ready for tiffin
> And wonderful fit;
> The morning round over—

A decent round too—
In the Nilgiri sunshine
Sheer gold out of blue.

They're old and they're past it?
By no means at all!
at the fourteenth Sir James
Drove a deuce of a ball,
And the Colonel's approaching
Was great, on my oath—
But the Nilgiri sunshine
Had a big hand in both.

The Nilgiri sunshine!
Sheer gold out of blue,
Heart-winning, heart-warming,
Heart-waking anew;
The flowers for its fairies,
The birds for its friends,
And its message—"Life's excellent;
Life never ends."

The sunshine of England
It shows in its way
That winter's not summer,
That night isn't day;
But it can't call the colours,
It hasn't the art
Of the Nilgiri sunshine
That kindles the heart.

Outside the Clubhouse
The sun blazes down
On the jolly old stagers
Rose-ruddy and brown;
Happy as sandboys
They sit there and thrive
In the Nilgiri sunshine—
That keeps 'em alive.

There's a breath in the wattle,
A stir in the gums;

Down his bright stairway
The sun-god comes—
Down to his Nilgiris,
Azure and gold,
Where nobody ever
Need fear growing old.[12]

Not only is the Nilgiri climate a relief from that of the Indian plains, we are told here, but it surpasses even that of England. As we saw in the earlier description of flowers and vegetables, the Nilgiris are represented as paradise: as England on a grander scale, a land blessed to excess by the nature gods. Those who reside there are blessed with immortality. The first two stanzas make it clear who these immortal residents are: it describes "the old fellows." They are, most likely, planters or foresters, sitting outside "the clubhouse"—a mandatory component of every British settlement, and a strategic cultural site for the reproduction of colonial ideologies—or playing cricket, their advanced age belied by the vigor with which they drive the ball. The poet would have us believe that they are uncannily fit for their age, somehow having escaped from natural aging processes such as might occur in less enchanted spaces, like England. English nature, the fourth stanza suggests, is far more prosaic, dull, decidedly unmagical, if functional and predictable. What gives Nilgiri nature its superior power? How can we understand this apparent inversion of hierarchies?

English nature, this stanza suggests, is part of a timetable that regulates the day and the year, dividing them into predictable elements, and producing temporal progression ("the seasons change, the years pass, men grow old"). In the Nilgiris, however, the Englishman escapes from the tyranny of regulated, ordered temporality into a space that stands outside time. Nilgiri sunshine has "the art" that English sunshine doesn't—this appeals to a view of art as outside of sociality and temporality, eternal, unspoilt by the griminess of production relations. Like pure art, then, Nilgiri nature is unburdened by social functions; its manifestations are fairylike, full of flowers and birds. The status of paradise accrues to the Nilgiris by virtue of the asociality of its nature. Naturally, then, when one escapes sociality and temporal progression, one attains immortality.

This pure uncorrupted nature and therefore this immortality are not available to everyone, however. There are at least two moments in the poem that allow us to unpack the self-understanding of this poem. Consider the flowers, fairylike, putatively sprung from a divine nature; it was no secret, either to the planters in the poem or to the author himself, how the flowers came to be there. The sweet peas and daffodils that adorned the landscape were cultivated by professional and amateur English gardeners, as we have already seen. The jolly celebration of nature in the first stanza is rendered unstable by the sociality that lies close to the surface of its purely natural elements. One only needs to tweak this strand of sociality, so to speak, to have a whole mess of power relations disrupt the happy surface of this celebration of nature.

Both metaphorically and materially, gardens were significant ideological spaces in British India. They were seen as bringing order and refinement to an alien and inhospitable land. In material terms, private and public gardens created spaces into which only certain types of citizens could legitimately enter: the civilized, genteel sophisticate with the capacity to contemplate nature in its pristine purity. Consider the word "nobody" in the penultimate line: Who does this exclude? By comparing narratives of Englishness with narratives of the native and his relation to nature, it becomes clear that this particular timeless construction of nature is available only to the highly cultured (that is, to truly appreciate nature in all its purity, one had first to possess the requisite cultural training or sophistication). This reification of a pure or elevated nature, along with its pair, an elevated cultural sophistication, was accomplished through contrast with a corresponding nature-culture pair, namely, low or debased nature and low or primitive culture.

One of the reasons indigenous worldviews occupied this lower position (defined by low nature and low culture) was because they were believed to have failed to separate the natural and the cultural; the natives had failed to distinguish between work and leisure, between the sacred and the secular. They had failed to make the transitions that modernity and industrial production required, whereby they ought to have separated themselves, as self-acting autonomous subjects, from their surroundings, in order to act truly upon a separately conceived nature. Instead, they confused nature and culture, ascribing subjectivity to natural objects (often naming hillocks, groves, and other features of the landscape) and refraining from that genuine productive activity which results from acting upon and transforming the landscape.

The ascription of subjectivity to natural objects is contrasted with the creation of a new and progressive society in the following description of the earliest European settlement in Ooty. "Lord Elphinstone, then Governor of Madras, fancied [a] spot for the erection of a dwelling house, but met with great difficulty in effecting the purchase of the ground, on account of the objections raised by the Badagas [the local tribal group], who had from time

FIGURE 5. Badaga Girls. From Edgar Thurston assisted by K. Rangachari, *Castes and Tribes of Southern India,* vol. 1 (1909; reprint New Delhi: Asian Educational Services, 1987), facing 75.

immemorial sacrificed a buffalo calf every year to a deity supposed to be present in an old decayed tree growing in that locality."[13] The narrator describes how the governor was able to persuade the Badagas, through gifts and drinks, to give up their rights to the land for a nominal fee. He continues, "No sooner was the transfer concluded than his Lordship began to enlarge the old building, and in course of time converted the property into one well worthy of a nobleman's residence. The house was magnificently furnished, . . . the grounds were tastefully laid out, and the whole assumed the appearance of a beautiful English Manor House—full of enchantment and attraction to the exiled Europeans, a perfect oasis in the surrounding waste."[14] Note the contrast between the impulse to acquire property and, through laboring, to transform it into a civilized land, and the premodern deification of space that leaves the land static and inhospitable. This contrast is sustained, for example, by the contrast between the "decayed" trees of the natives and the "enchanted" oasis of the Europeans.

Despite the implication in this passage that indigenous ties to the land were easily swept aside, it was not the case that all Nilgiri tribes acquiesced passively to either the appropriation of their land or the succeeding upliftment programs of the state and the church. In a later section I touch briefly on the complex issue of indigenous resistance and the competing narratives of nature offered to the English settlers.

Power and Authority

Why did the colonizing group see itself as somehow fit to take charge of the land? The form such justification takes is identical to arguments made for the displacement of almost every indigenous group displaced by European colonists between the fifteenth and nineteenth centuries. Elizabeth Povinelli reminds us that "early political-economic theory postulated that laboring subjects created proprietary interests in things and that the mode of production determined the level of those proprietary interests. And colonial law settled Australia as *terra nullius* based on this assumption."[15] In other words, since Australian aboriginals did not labor over their land, they were not entitled to own the land they inhabited.

The British could never claim the *terra nullius* argument in India as a whole, since precolonial Indian society was well known as being constituted by complex and diverse modes of production, and as having its own elaborate land tenure systems. In tribal areas, however, most often in hilly forested regions such as the Nilgiris, an argument very close to this was implicitly made. Categorizing vast tracts of hill land (long used by tribal populations) as wasteland or unused land, the government initiated forest management programs and encouraged coffee and tea planters to bring the land under modern productive systems.

There is a considerable body of historical scholarship on land tenure in precolonial and British India, which I do not attempt to synthesize here.[16] Once again, however, popular narratives are invaluable windows into the manner in which ideological constructs of the economy found their way into commonsense thinking. Consider the following extract from an anecdotal memoir of an Englishman who had lived in tribal hill regions for most of his professional life. The author is an avid hunter and nature lover of the contemplative kind found in the verse we read earlier. He has been part of a half-century of hill development, and comments here on the requirements of civilizational progress and the incapacity of the native hillman to fathom them. After extolling the simplicity and honesty of the hill dwellers, he goes on to explain:

> [A]nimals have no money and no need for it; and it is much the same with the villagers of the Agency whose lives have been set by a kind and merciful Providence very nearly upon the animal level. . . . [T]he characteristic which most distinguishes the Agency man from his fellows is a hopeless and absolute apathy. He does not want to do anything, he does not care about anything, he does not value or seek or aspire to anything—which is a shade trying when your ostensible object is the reformation of him and his country. He does not much mind whether he lives or dies. [I] ha[ve] seen an Agency coolie go down with heart failure after carrying a heavy load of mangoes up a ghat, and the man crawled to a pool of water and lay down and died aimlessly in the most disinterested and casual way. . . . With this example in view you will not expect to hear that the Agency peoples show a quick interest in clearing jungle for cultivation, building roads or setting up sanitary houses for themselves or others; and indeed they do not. . . . How it is that [they do] not starve or die of sheer inertia is a problem science has yet to solve.[17]

Lockean assumptions regarding the constitutive link between personhood and private property enable this narrative to take as self-evident the impossibility of progress for the native.

The author of this narrative also gives us a glimpse into the significance of hill-station life for the colonial administrator who chafed under the bureaucratic restrictions of life in the plains, where he was more directly responsible to government scrutiny (British administration was known for its complicated bureaucratic procedures). Thinking back to the poem about Nilgiri sunshine, one can see now that the frisson of immortality experienced by the happy cricket players came not merely from the effects of a particularly sunny Ooty afternoon; it was, in fact, a consequence of the increased power and authority wielded by a colonial official in the hills, as compared to the plains. Take this account of official duties in the hills: "what a godsend and relief it is to get a little elbow-room, a little broad freehand work, a chance of action not absolutely determined by Codes and Manuals. . . . Oh, the relief of getting away [from the plains;] from niggling and pottering over Section X Subsection Z to a wider world where one lays out roads and reservoirs, plans for reserves and areas for cultivation, settles questions of rights and tenures, has, in a word, a chance of getting something done that shows."[18]

Officials in the hills could easily see themselves as invested with godlike qualities: a sense of omnipotence emerges, in this account, from the language of laying out, settling, and planning the lives and landscapes of the hills. Hill stations were relatively isolated from the state bureaucracy, and hence individual administrators were less accountable to higher officials. This increased personal power allowed hill administrators to feel like lords of their own kingdoms.

In moving from discussing climate and vegetation to discussing power and authority, we have not really moved outside of an environmental narrative. Deconstructing constructions of nature offers a supple route to political analysis. Our notion of environment ought to encompass social and political dynamics as well as the physical landscape, since the two continually constitute each other.

This brings us, of course, to the question that has been in the background since the beginning of this discussion: whose rights and tenures were being thus authoritatively determined? Who inhabited this land of Eden before it was discovered to be a paradise? The answer is no secret. The Nilgiris, in addition to being heaven on earth for British officials and their families, were an ethnographer's paradise, and scores of anthropologists, amateur and professional, flocked throughout the century to record what they considered valuable data on the origins of man, studying various aspects of the five tribal groups of the Nilgiris.[19] The narratives generated by European encounters with these groups will offer us another perspective on colonial constructions of the environment, and on the constitution of the "hill tribes" as anthropological objects.

Natives

In his account of the early European settlement in Ooty, the German missionary Friedrich Metz describes the disenfranchisement of the Badagas from their land, in a passage we saw earlier. He describes how the land was obtained at a nominal price from the Badagas: "It used to be the boast of the old headman that the Government once came in person to ask for the site, and that he maintained his rights against him. It is said that what His Lordship did not accomplish, was afterwards secured by his Lordship's steward, who feted the Badaga chiefs and when he had got them into good humour, persuaded them to give up the land on condition of receiving an annual fee of 35 Rupees. The objections of those who had a prescriptive right to the soil being thus removed, his Lordship obtained the land on a lease for 99 years."[20] Here we see the same land of enchantment invoked, but this time at the end of a narrative that exposes the other side of the bargain. Nor were the new inhabitants content with having gained control of the land; they had grander plans in mind: environmental and social progress through systematic planting, efficient production, and a pedagogy of scientific progress.

Pedagogical efforts to uplift the natives were regularly made by missionaries, planters, and forest officials. In 1845 Elphinstone's house was bought by Mr. Casamajor of the Indian Civil Service, who established a school for Badaga children. Metz comments: "but so little were the hill tribes able to appreciate the value of mental culture that it was necessary to offer them a douceur of one anna per diem for every child sent to the school."[21] Metz himself spent several years trying to uplift the Todas. Sullivan and administrators who succeeded him appointed chiefs of the Todas to oversee the communication and authorize agreements between them and the Europeans. These appointments resulted, over time, in the undermining of traditional forms of authority. The Toda economy revolved around buffaloes and milk; their religious authority was known as the Pálaul, the person who oversaw the care and worship of the buffaloes. Metz comments, "Prior to the period when Europeans began to resort to the Neilgherries, the authority of the Pálaul must have been very great, but of late years it has much diminished; and as the light of civilization has penetrated into the dark corners of the Hills, superstition has quailed under its influence; and even the self-righteous ascetic in his secluded retreat has gradually relaxed in those rigorous austerities which the religious notions of more primitive times had imposed upon him."[22]

Several ethnographers noted that the Toda birth rate began to decrease following European settlement of the Nilgiris; some noted a rising incidence of venereal disease and sterility. Many ethnographers expressed the concern that the extinction of the Todas would mean the loss of valuable data. In Metz's view, "one thing alone can ward off this result [extinction], and that is a hearty reception of the Gospel with all its purifying and elevating effects upon the life and habits of those who embrace it in its fullness."[23]

Christian missionaries were moderately successful in gaining Toda and Badaga converts; in most cases the converted communities lost ties with their original kin and social groups. However, it was not the case that all Nilgiri tribes acquiesced passively to missionaries' and revenue officials' schemes of progress. Through lies, threats, and the possibility of flight, they carried on what were ultimately unsuccessful but often vigorous efforts at resistance.

Reading Narratives of Resistance

Implicit in the historiographical act of rereading colonial romanticist narratives of the land as discourses of power is the assumption that one can read hegemonic narratives against the grain, in order to study the forces, both dominant and subordinated, that formed the agonistic domain in which these narratives were forged. Reconstructing a subordinated narrative so as to restore historical agency to the marginalized is often seen as the next step in this counterhegemonic historiography. Such a process was pioneered by the Subaltern Studies collective, whose many successes have nevertheless been accompanied by vigorous debate over the precise nature of their liberatory effects.[24]

The critical framework of a cultural studies of colonial science can unravel narratives of scientific progress so that the constructedness of a putatively seamless scientific rationality is revealed. The excavation of subordinated narratives of science will be far more fraught with difficulties than the historiography of, say, agrarian protest, simply because the hegemony of scientific ideologies has historically been more insidious, and therefore less contested, than the politics of land revenue or labor laws.[25] There are, however, several points at which the colonial narratives of nature in the Nilgiris reveal fissures through which the agency of the indigenous inhabitants clearly asserts itself.[26]

Lies and Evasion

Friedrich Metz, the anthropologist-missionary who sought to save the souls of the indigenous Nilgiri tribes, offers us narrative histories that clearly reveal the difficulties of assigning passive, backward, nonrational subject positions to these unsaved souls. Even as Metz anxiously asserts his own readings, the historian is offered generous glimpses into the contested terrain in which such assertions had to be made.

Metz tells us, for instance, directly and rather indignantly, that Toda informants have long been lying to European inquirers: "A familiar acquaintance . . . with their language . . . ha[s] led me to discover that much dissimulation is practised by these men towards Europeans, and that they soon detect what information the latter desire to obtain and make their replies accordingly. The custom of paying the Todas for every insignificant item of information has naturally brought about this result, and it is now a matter of difficulty to obtain from them any account of their previous history, upon the truth of which implicit reliance can be placed."[27]

It is evident from Metz's account that the Todas were by no means conveniently stable objects of scientific investigation; processes of construction were evidently happening on both sides of the colonial encounter. The Todas, gauging that these strangers who tramped with increasingly regularity across their pastures would pay sums of money for information about Toda history and customs, obligingly invented stories according to the context in which the question was asked. Far from being evidence of their fundamental irrationality or incapacity to think historically, as was postulated by frustrated enquirers such as Metz (who was one of the few who had, in fact, spent enough time with the Todas to notice such discrepancies), this was probably due to the fact that anthropologists' questions were couched in terms of their theories of progression or degeneration of races, and were unreflexive about the ways in which they themselves were perceived as interlocutors. Toda responses, then, were rational responses to a situation in which they were presented with questions whose formulations did not pay attention to the interlinking of their histories, landscapes, and social practices; and, moreover, appeared to be attached to a reward of which they might avail themselves by satisfying the questioner with any combination of narratives.

In one of the Toda funeral songs recorded by Metz, different forms of sin are enumerated. Among these are killing a snake, lizard, or frog; drinking water at a stream without first worshiping it; destroying the walls of a water reservoir; and making a complaint to the government. We might note the environmental utility of these practices. Respecting sources of water, protecting useful wildlife, and evading government supervision are accorded equal value. The inclusion of the last into a funeral song indicates that the strategic value of lying to colonial officials was a tactic that was being written into the society's normative code of ethical conduct, since the funeral was one of the occasions for the production of rituals that reproduced social and religious norms and reinforced social cohesion.

Hostility and Threatened Flight

Metz relates that although he was initially seen as nothing but a harmless storyteller, when his intent to proselytize became known there was open hostility. Badagas refused him admittance into a village; in a Toda village they refused to give him milk, saying that the buffaloes would be destroyed by offended deities.

When the government issued orders restricting Toda buffalo sacrifices, Metz records that "When this prohibitory order was issued the Todas charged me with having been the cause of it, and vented their maledictions upon me. . . . [They] threatened to flee from the hills, but I declared myself prepared to follow them wherever they might go; and in course of time they have become reconciled to the restrictions placed upon the mode of observing their ancestral customs." [28]

Metz also refers to the "hill brahmins" or Haruvaru, who would walk on coals. Metz scoffs at "so palpable an imposture." He suggests that "they remained only a few seconds on the coal" that there was no possibility of burning. "I used often however to be taunted by them and challenged to perform a similar miracle in the name of my God: to which I replied that a much greater wonder than that was that I should not cease to love such squalid fellows as they were; and that too in spite of all the abuse they daily heaped on me. When a recent order was issued prohibiting the practice . . . I had to submit to the most dreadful execrations, showered upon me as the originator of the order."[29]

It is evident from Metz's own narrative, thus, that there was vigorous resistance to the imposition of a Christian narrative of the environment and humans' place in it. The Todas, represented in anthropological treatises as a passive, peace-loving tribe, lethargic to the point of eternal indolence, were not incapable of causing the intruding missionary to flee when he became too presumptuous.

Other Narratives?

It is possible to guess at a narrative of the Nilgiri environment that is quite different from that the English Promised-Land narratives I have described so far. From the reports of anthropologists and missionaries, a picture emerges of a landscape invested with social and religious significance by the different groups that inhabited and used it. Metz reports in his tirade against idolatry that there were at least 338 Toda idols in the surrounding area, placed on hills, trees, or by streams. Each natural feature of the landscape had a deity associated with it: Gangamma was present in every stream; Bété Swamy, the god of sport, had a particularly large tree as his resort, and Mukurty Peak was named after a piece of a demon's nose. Metz complains that "a host of nameless gods . . . live in the heart of the lofty mountains, and in dense impenetrable forests."[30] Landscape narratives clearly conflicted here: the colonial narratives, while steeped in romanticism, saw themselves as securely anchored in a rational, resource-oriented scientific and technological paradigm. The indigenous narratives, which combined the conceptualization of nature as resource with that of nature as socioreligious formation, were accordingly constructed as the Other of rational scientific discourse.

Interestingly, a later missionary, Catherine Ling, who spent forty-five years in the Nilgiris and became a trusted member of Christian Toda communities, was also written into the landscape, or had the landscape written into her in the following Toda song.

> She has become with her steps this way on the way.
> She has become with her nose this way on the path.
> Moss has risen on the branches. A hump (lit. mushroom) has risen
> on the back.
> She has become like the sun setting on the hilltops.
> She has become like a tree yielding ripe fruit on the high branches.[31]

We can see that the Todas commonly exercised a discriminating taste when assessing the intentions of visiting colonialists; they were far less likely than the colonizers, it seems, to indulge in the creation of monolithic discourses of the Other.

Whereas amateur Nilgiri poets and popular writers might use religious metaphors of the sun god or bountiful nature to extoll the beauties of Ootacamund, their narrative of the land sits firmly within a post-Enlightenment secular worldview, by which a romantic contemplation of the landscape coexisted with a model of utilizing it through systematic and productive "improvement." In the next section, I examine this narrative of improvement by expanding our archival exploration a little farther than the confines of the "English" town of Ootacamund. Narratives of nature were not, of course, confined to concerns about hedgerows and daffodils in Ooty lanes and fields. As examples of the routes by which the examination of nature narratives rapidly takes us through a number of different discursive fields, I will briefly touch on the discourses of hunting, planting, forestry, and health.

The Scope of Nature Narratives

Hunting

The narratives of hunting, or shikar, offer the historian a host of insights into this romantic yet masculinist and progressivist construction of nature.

> It is a mistake to think that people are fond of shikar because they are fond of killing beasts. . . . [My] idea of shikar is to go out into the shikar country and meditate. [I have] . . . sometimes thought that if one can absolutely overcome one's body so that it becomes a mere block of substance, absolutely silent and absolutely still, one's spirit can get away from it and come into tune and touch with the place around one as do, no doubt, the spirits of animals . . . the inducing of a certain peacefulness of mind and a tantalizing sense of getting nearer to the heart of things.[32]

Here we have the narrative of a romantic communion with nature that all shikar stories include. Noting the language of metaphysics, one might ask what it is that distinguishes the romantic-religious narrative of nature here from the spirit-imbued Toda narratives decried by Metz. We begin to understand how this line is drawn when we note that what is left out in the shikar narrative above is the indigenous expertise and manpower that provide the material basis of this metaphysical experience. But we do not have to look far, for the British shikari always mentions his native assistants.

> Always, of course, one has one or more native shikaris; these are to be distinguished from their fellows by their extreme hideousness and by a skinniness which approaches towards hallucination. . . . Yet it takes a man in the very pink of condition and training to keep up with them on a long day's hunt. . . . Their capacity for work seems to increase slightly the more hideous and emaciated they are. . . . For all this, however, it is these wretches, armed with a condemned Police carbine or an old double-barrel or a country musket of a type unknown to modern makers, who kill off the game and spoil the Indian shikar.[33]

This narrative suggests that the practice of killing animals with guns is not quite the same when practiced by emaciated natives as by serious English hunters in khaki shorts and solar topis. It assumes that the romantic appreciation of the wilderness is accessible only to a culture that has transcended the brute materiality of hunger and labor. This transcendence, of course, occurs via scientific forms of improvement.

To show how the shikari's romanticist view of nature went hand-in-hand with a pragmatic, scientistic approach to improving nature, let us look at some narratives of systematic planting and of productiveness applied to forests.

Planting

Intersecting narratives of afforestation, deforestation, and forest labor were generated around the Nilgiris. To discuss this we must move to the northern part of Nilgiri District (the Nilgiri Wynad), more densely forested than the region around Ooty, and thus the focus of attention for the forest department. Some of the material here was generated in relation to Wynad proper, part of Malabar District. I include this, and some comments on other parts of the Western Ghats, as these were contiguous forest belts and part of the same mountain range (the Western Ghats) and also of the same forest management system, under the Forest Department of Madras Presidency.

By 1856, when military barracks were being built just outside Ooty, H. R. Morgan, deputy conservator of forests, noted that there were no long logs to be had in Ootacamund. The Forest Department thus obtained "a rented forest adjoining the Wynad Teak Belt and by 1860 delivered 200,000 cubic feet of timber to the barracks at a cost of only 1 rupee/cubic foot, after all charges had been paid."[34] The labor he employed for this job was recruited from among the Kurumbas, and here we can detect conflicting narratives of the environment. "In working this forest, . . . the Coorumbers had to be trained to fell only the largest and best trees, as they were in the habit of felling a tree that would just measure about 12 cubic feet, as they were paid by the log. I introduced the system of payment by the cubic foot. . . . At one time they gave much trouble, every large tree was 'Swami Tree' and could not be cut; even this difficulty was surmounted."[35]

Here the native is represented as a good source of labor, if firmly supervised. The trouble here is caused by the natives' refusal to cut trees; later we shall see that trouble also arises when natives do cut trees. For timber operations, Morgan soon developed a typology of labor.

> In Wynaad the axemen are Corumbers. . . . They are also very useful for Teak plantations, many are intelligent, and the great advantage of employing them is that they live in the forest all year round, they fell and square timber with great precision, they can also be trusted in planting out operations; for cartmen, road labor, &c., Canarese are employed. Sawyers are obtained from Mysore and other parts. On the Anamallies, men from Palghaut are employed as axemen, they were very expert in dividing by means of wedges very large trees into planks suitable for dockyard purposes. The Kadirs, a jungle tribe, are useful for building huts, the Mahouts there and in Wynaad are generally musselmen, whereas in Nellumbore and those parts, they are almost invariably Punniars, and as the Nellumbore elephants are used without harness, dragging by their teeth, the equipment of a Punniar and his elephant may be said to amount to *nil*. These Punniars obtain an extraordinary influence over their elephants, and especially the males by a peculiar process. The local labor at Nellumbore is made up of Malayalums and some Moplahs, and there are many trained men amongst them who understand planting and pruning.[36]

We note here that the forester was expected to build a detailed ethnographic understanding of the "inherent" capacities of various tribes. The responsibility of a forest officer, ran the dominant narrative of the Forest Department, was to conserve the forest against wasteful use and to harvest the forest for progressive and profitable use. We can see this in Morgan's textbook: "It has always appeared to me that if the Forest Officer had an efficient establishment, his duty was to carry his timber to the best market." And, further, "the ryot should be saved from his own destructive habits and taught that to destroy forests is not the way to benefit himself, and that by a little timely forethought, he might procure forage for his cattle without having recourse to the reckless system pursued by him for ages."[37]

In order to fulfill his duties competently, a forest officer had to have multiple talents.

> [A] Forest Officer to be really useful should know first and foremost the language of the people, without this he is useless, or at best, at the mercy of intriguing interpreters. In addition to the language, the habits and customs of the people, then Arboriculture in all its branches, next Engineering and Surveying—how to build houses and bridges, survey roads, and blocks of forest, &c., run boundaries by

> a pickaxe trench, as mounds of earth, stones are useless. Physic his people when ill, treat them with tact, attend to the health of his bullocks and elephants, and last but not least, keep his own health. The life of a Forest Officer is not cast on a bed of roses, but rather a bed of thorns, an iron constitution and a good conscience may enable him to surmount all his difficulties.[38]

Here we are told that the Forest Officer had to be an expert in all sciences—especially engineering, ethnography, and medicine. This knowledge would confer on him the power to be nurse and protector to both nature and human. The narrative of expert knowledge leads us to the narrative of paternal protection.

> In vain does the Forest Department try to save the ryots [cultivators] from themselves, their improvident habits too surely destroy the jungles which ought to last them for centuries. . . . It is not of the slightest use attempting to raise the status of the ryot, so long as we leave the main points of his position untouched; and until he is educated up to a certain point, he is quite incapable of appreciating any efforts made in his favour. It may be urged that this has nothing to do with forestry. I maintain it has, for we have to consider communal rights and to teach the ryot not to encroach on reserved Government forests. . . . It is to be borne in mind that the ryot has for ages been brought up under a reckless system of mismanagement, and new habits are not learnt in a day especially by a people who are the most conservative that the world has seen. From sire to son for ages, the same habits, trades, occupations have been the hard and fast rule. . . . [T]he eternal laziness of the ryot prevents him from ever attempting to work for his descendants.[39]

Here Morgan clearly lays out the paternal role of the forester: he must not only manage the chaotic tropical forest, but also drag the resisting forest cultivators into the modern economy. The narrative of paternal instruction in the care of the environment found an ideal candidate for the villain of the piece in *kumri* cultivation, otherwise referred to as "slash and burn" agriculture. Morgan decries it: "Of all the varieties of cultivation carried on in the world, this is the most pernicious." He lays out plans for its curtailment.

> The aboriginal tribes must be found some other means of employment. They must be civilised. . . . They must be encouraged to plant

> fruit trees and grow other crops than those they have been accustomed to. They must be taught trades. . . . Congenial employment must be found for them by the Forest Department, such as timber squaring, the collection of forest produce, &c., and deserving men should be employed as watchers, elephant-drivers and maistries [overseers]. They can be made of the greatest service to the Forest Department or the reverse; as they happen to be treated. . . . How can the Curumber be expected to rise in the scale of civilization when he dare not possess property?[40]

We come full circle, then, to where we started: the route to abundance is through the planting of "fruit trees" and crops other than those indigenous to the area; progress can only occur if respect for private property obtains. In Ooty, we had another group of people formulating property rules and planting English fruits and flowers.

We can read these narratives not only for the narrators' attitudes toward the land, but also for the political values they advocate and the power relations within which they are located. The story of interlocking environmental narratives is by no means completed with the elements described above. It was not only the government that cleared and planted forests but also private entrepreneurs who cultivated the plantation crops (rubber, tea, coffee, and cinchona) for export to European markets. There are more strands in the narratives of nature. Let us consider, briefly, the medical discourse, the private planter's discourse, and the international political economy of transplantation.

Health, Labor, and Planting

In the 1860s, there was an anxious medical debate over the possible correlation between kumri and malaria. Reports from the Madras Government Medical Department, and letters between the government, the district collector, and *tahsildars* (local revenue collectors) reveal several different speculations about the nature of this correlation. One of the most interesting was expressed by Cole, the inspector general of the Medical Department, in 1865. "I quite agree with those gentlemen who think that the soil of the Kumari [sic] cultivation is much more likely to be a source of fever than when covered by its original jungle. Whatever the ultimate essence of malaria may be, we know that it is largely contained in the soil, and that the first disturbance of the surface soil, whether for cultivation or other purposes, is often followed by outbursts of fever."[41]

He thus suggests that the clearing of the forest that is carried out by tribals such as the Kurumbas is an unhealthy practice and might breed malaria. He thinks, however, that it is a good idea in general to clear the tropical jungle: "As to the general question of permanently clearing and cultivating spaces in the dense forests of the Western Ghats, there can be no question but that such operations here, as in every other portion of the world, must ultimately result in sanitary benefit to the population. Excess of vegetation is, and probably will always be, in any part of the world, incompatible with the numerical increase of the human race."[42]

What is the significance and importance of clearing the forest, then? In the private planting discourse, clearing the forest for cultivation is seen as a positively valued activity, one that aids the progress of civilization. An anxiously repeated if not always coherent distinction is constantly made, however, regarding the difference between progressive, scientific European cultivation of the land and unscientific native forest clearance.

> The feeble races of N. Canara can make no headway against the jungle. If the latter is to be in part reclaimed, it will have to be done by other than indigenous labour. There can be no question, in a sanitary point of view, of the benefit to these wild districts of encouraging Europeans of means and ability to take up and permanently clear patches of these vast forests, such as may be suitable for coffee, tea, or cinchona.
>
> In fact, until very considerable portions of the Ghauts are cleared, it is very doubtful whether the climate will ever permit of the districts being inhabited by a healthy and industrious people.[43]

The argument here suggests that forest clearing must be done under the supervision of Europeans, and must be for the purpose of planting crops other than the kumri crops, paddy and millet. It is only the plantation crops that truly bear the potential for the liberation of the land from its unhealthy state and of the population from its parochial agricultural practices. In an 1864 handbook for coffee planting, we learn that "Everything on a plantation should be carried out scientifically, order and regularity being studied in all cases, no matter how trifling they may appear. The beauty of the plantation should be looked to as well as the productiveness. . . . [S]traight lines . . . render the plantation regular."[44]

We learn from another planting memoir that planting in the Nilgiris has many advantages: the local tribal labor is cheap, and extra labor can easily be shipped in from neighboring districts, in which unemployment is high.

Further, the planters in this district have successfully lobbied the government to build public roads connecting them to markets and to urban machinery and supply centers.[45]

However, we are reminded that planting is a serious business, not one that is available to any ignorant agriculturist.

> The natives . . . are daily opening their eyes to the profits derived from coffee cultivation, and it is not uncommon to see their threshing floors for rice filled with coffee. . . . Little, however, can be said in favor of these consumers of betel mixture—whether it is that the mildness of the climate and fertility of the soil render active exertion unnecessary, but these people seem to regard *sloth* as the chief luxury. Jungles have risen to an enormous price. . . . Some . . . Natives are insane enough to suppose that if they put in seed coffee on the ground, regardless of the wood or underwood being cut, that they will be in such a case entitled equally with the European planter to hold their jungles as coffee plantations; but surely Government won't be taken in by such a silly trick. Almost daily, exceedingly fine offers for jungles are made by Europeans to these [natives], but they will accept nothing in reason. . . . It is altogether a dog in the manger case.[46]

The argument being made here, evidently, is that the natives are perhaps subject to a form of climatically determined activity disorder that makes it impossible for them to be true planters; to plant meant that one proceeded in an orderly fashion; one hired labor to clear the jungle and to plant in straight lines; further, one had to avail oneself of roads, communications, and the economic infrastructure that connected producers to markets. The natives could not, by definition, qualify for this position.

This narrative is incomplete if it ends with yet another local discourse of the Nilgiri Hills. This space in south India was itself never purely local; although I have been piling on the local detail, it should be emphasized that my categories of analysis are in fact globally significant in this historical period. British colonial notions of scientific progress were rooted in a system of thought that was neither universal nor transhistorical. It employed sets of oppositions between nature and culture, work and leisure, labor and sloth, active and passive, high and low culture, that were specific to the political and economic context of England after the Industrial Revolution. E. P. Thompson and Raymond Williams have shown us how these oppositions were ideologically employed in similar ways with respect to the transformation of the English peasantry and the creation of a working class.[47]

To understand the political economy of such a narrative, finally, one must move even farther out from the Nilgiris, to examine the global botanical networks that linked colonial peripheries together through the center of botanic expertise, Kew Gardens. In the mid-nineteenth century, the British government launched an ambitious project, centered at Kew, to manufacture a cheap drug against malaria, which was claiming thousands of English lives and posing a danger to military strength and public health. The production of quinine (the antimalarial drug extracted from the cinchona tree) resulted in the establishment of an enormous global network of exploration, collection, and systematization of botanical knowledge, a centralized array of botanical gardens, and a colonial science of natural resource management. This is a narrative that offers yet another perspective on the questions I have raised here: how did economic and political concerns inform theories of how to manage nature? What distinguished local, indigenous ecological knowledge from European botanical knowledge of the same physical environment? Chapter 7 ponders these questions, and offers the cinchona story as a brief historical moment that had far-reaching consequences. Before we reach chapter 7, however, we will prepare for these questions by investigating some scientific and nonscientific practices that altered the lives and landscapes of local communities who lived in and around the forests of Madras Presidency.

CHAPTER 3

Forests

I'm travelling on a wild-goose chase,
Trying to frighten each native race,
Musselmen, Hindoos, Christians, and what not,
The last, (Roman Catholics,) the worst of the lot. . . .
The state of the timber is bad, and corrupted;
But there's a rich crop of natural produce,
such as gall-nuts and soap-nuts, and others of great use;
Besides certain roots, such as saffron and ginger,
And a few of those spices that season a swinger.
Now the natives (in and without the pale)
These natural products steal wholesale.
I'm trying my best this year to outwit them;
But, as for the products: I wish I may get them.[1]

By the end of the first quarter of the twentieth century, the Forest Department could report its satisfaction with native "cooperation," despite the fact that the subcontinent was burning with noncooperation. The 1924 *Review of Forest Administration in Madras* reported that "The Government consider that the new policy of enlisting the cooperation of the very class of person which traditionally used to regard the Forest Department as its enemy, in the management of the smaller reserves has already had a most beneficial effect not only in reducing the possibilities of friction, but also in educating public opinion to appreciate the necessity of scientific protection."[2]

Former "enemies" of the forest, the native tribes, had been enlisted as friends. Public opinion had been educated to appreciate the necessity of scientific protection. Even after correcting for the requisite self-congratulatory tone of administrative reports at a time of growing nationalist challenge to the institutions of the British Raj, we can see this report as symptomatic of a momentous change in looking at nature.

The change in attitudes toward nature was neither a cultural shift from tradition to modernity nor a rhetorical shift from harmony to control, as many critiques of post-Enlightenment science have suggested. It was part of a shift in modes of production and representation, through a process in which forests were redefined as factories, nature as manageable resource, and forest products as commodities. In this process, the categories of labor, production, and property were key. The appearance of these categories was not inevitable or natural; we can trace the coercive discourses and practices that accompanied their naturalization. The new structures of representation redefined personhood through property, identity through labor, and progress through the imperatives of global production.

Teaching Respect for Trees: Whose Use Is Good Use?

Scratch the surface of the Forest Administration reports, and they quickly reveal the constant conflict that lay below forestry's public success story. Memos, letters, and arguments among forest officers and subordinates paint a picture of a staff earnestly and authoritatively instilling a whole new way of seeing the forest through managing the trees.

Complaining about children of the Chenchu tribe (whose home was in the northern forests of Madras Presidency), the forest conservator grumbled in 1928 that they had "been a menace to the forest—and they must be taught from infancy that their interests are bound up with the forest." He recommended that the children be put to work planting and tending, in order to learn "respect for trees."[3]

What was the world coming to, if Chenchu children had to be taught respect for trees? The world was quite literally being transformed by the discourse and practice of forestry, and its contradictory statements suggest where the fault lines of the new ideology lay. Conflict arose around "respect for trees" precisely because of the difficulty of naturalizing a new ideology: one that sought to inextricably link a good upbringing with ordered production, sound identity with unquestioning devotion to labor. In 1930, the *Proceedings of the Chief Conservator of Forests* reported, "As usual the members of jungle tribes continued to be protected and employed by the department. . . . They are everywhere largely dependent upon the department which treats them fairly and realizes that they can be made useful assets."[4]

But despite the benevolence of the Forest Department, their jungle wards continued to be uncooperative. The trinity of wage labor, profitable production, and private property seemed no more natural to the locals than it

FIGURE 6. Kadir Treeclimbing. From Edgar Thurston assisted by K. Rangachari, *Castes and Tribes of Southern India,* vol. 3 (1909; reprint New Delhi: Asian Educational Services, 1987), facing 176.

had a century before. In the Upper Godavari forests, home to the Koyas, "tribes were reported to be disinclined to undertake the labour of clearing the land"; plans for *taungya* had to be abandoned "as it was only by coaxing and pressure that work had to be got out of the Koyas." The report complained that the Koyas did not want to cut trees. "The Koya dislikes clear-felling these coupes because the trees are much bigger than those he finds in his own *podu* . . . [where] he leaves very big trees which do not dry quickly *eg. Odina, Sterculia* and *Boswellia*. . . . This system [*taungya*] will be changed into one of clear-felling followed by departmental burning and artificial regeneration."[5] Unable to teach the Koya tribe that respect for trees required their felling, to be followed by crop cultivation, the Forest Department had to resort that year to burning the area and regenerating a usable timber commodity.

Does this reluctance to fell trees indicate an inherent opposition to the destruction of nature in tribal identity? Far from it. The Chenchus, like many other forest groups, distilled a liquor from the flowers of the ippa (*Bassia latifolia*, or mohwa) tree. In order to collect the flowers more easily, the Chenchus would set fire to selected portions of the forests, which enraged the foresters. Further, once having collected and fermented the flowers, they would become drunk and "very unmanageable," commonly assaulting forest officers when in this state.[6] In 1876, an *Indian Forester* report noted that "The pastoral races annually fire the grass and undershrubs. . . . It should be sternly repressed."[7]

This *Indian Forester* article commented on the need to put forest production on a scientific footing, citing the recent scientific debates on the effect of trees on climate, and reporting on a House of Commons report entitled "The Effect of Trees on the Climate and Productiveness of a Country."[8] The *Indian Forester*'s account of this report allows us to see clearly the link between the environmentalist agenda of the foresters and the departmental attitude toward local users of the forest: "It will be necessary to exclude wood-cutters, cattle, sheep and goats from tracts allotted for the growth of trees. . . . Darwin strongly urges the need for preventing cropping by cattle, and for enclosing tracts on which it is desired that trees should grow."[9] The foresters thus aligned themselves with science as protectors of nature, casting the locals in the role of destroyers of nature. Forests all over the subcontinent were the homes of tribal groups who, by hunting and gathering or by clearing patches of the forest for crops, depended on the "destruction" of the forest for their livelihood; further, there were villages that depended on forests for their firewood and grazing needs.

In the mid-nineteenth century, faced with the need to manage both forests and their inhabitants, foresters had arrived at an efficient solution: to enlist the local users in their scientific forestry agenda. In addition to solving the problem of competing claims on the same resources, this had the added advantage of supplying the foresters with a pool of very knowledgeable labor, as the forest tribes had knowledge of and access to the forest at a level of detail that no one else, least of all the European foresters, possessed. Enlisting forest tribes in the scientific forestry agenda could not, however, be phrased in terms of a partnership in a scientific endeavor; it had to be incorporated into a morally inflected narrative of scientific progression, analogous to the one into which nature itself had been incorporated in the official discourse of forestry. This was made easy, indeed natural, by the existence of numerous theories of human advancement (most strongly propounded, in this context, by anthropologists) that placed wandering tribes below settled cultivators, and those below industrial societies, in a theory of linear and inevitable human progression. The key factors in catalyzing the succession from one stage to the next were private property and science. The *Indian Forester* report recommended: "Non-Aryan races are field labourers and should be made farmers and holders. . . . Besides the settled non-Aryans, there are, in all parts of India, numbers of homeless races; in the Tamil, Telugu, and Kanarese countries, not fewer than 100,000. Their history cannot be traced, but they are contributing nothing for the support of themselves and others; indeed, most of them are predatory. . . . Tact and patience will be required to withdraw them from their migratory habits, and induce them to settle down to agricultural pursuits and become producers."[10]

Although property was the material requirement for progress, the legitimating factor was, as we can see in this quote, production. Production functions here as an assumed marker of advancement: the only way in which wandering tribes might insert themselves into history is through settling down to systematic agrarian production. The presence of science as a guiding factor through this reasoning helped to present such a model as an interest-free, positively valued model of progress.

If we attempt to get behind this model's smooth surface and ask what interests might have constructed and sustained it, it quickly becomes clear that economic, moral, and scientific arguments are closely intertwined. If the colonies, and specifically India, were to function efficiently as sources of raw material for metropolitan commercial and industrial needs, it was essential that all their disparate modes of production be drawn into an integrated system of production for distant markets. Both global and national markets had to be supplied, and unless every able-bodied colonial subject produced his share, the markets could not function at their optimum capacities.

In the case of forests especially, two factors made it imperative to enlist the forest tribes in production for global markets. First was the problem of accessibility. At mid-century, transport and communication systems in hilly and forested areas were not yet developed. Early foresters were dependent on locals as guides into the forest, as aides in identifying species and their uses, and as axemen for clearing or felling timber. A consequence of this was, in the later part of the century, the foresters' dependence on these same local communities as labor pools for the continued functioning of now-established timber (monoculture) plantations. Second was the problem of forest use and capacity. The forests had already been actively used for centuries by local groups; once colonial production brought to bear the needs of global markets, however, the forests could no longer sustain the intensity of use. This led to the introduction of the scientific system of forest conservation after about fifty years of indiscriminate felling for shipbuilding, construction, and railway needs; further, it also led to drawing up plans to limit local use of the forest to those activities which would not cut into imperial profits. The solutions to both these problems (local knowledge and competition for resources) could be nicely combined by settling tribal groups in locations conveniently placed within official "working circles" (the term used for the officially surveyed and demarcated administrative units into which forest areas were divided), thereby creating a predictable and controllable source of labor with the specific skills required for everyday forest management work.

Forestry in Madras: A Brief Overview

E. P. Stebbing, looking back on the idea of the forest village, wrote as follows in 1921.

> It was in India that the idea of the forest village was, I think, first developed. Difficulties had been experienced in obtaining the necessary amount of labour to carry out the annually recurring work in the forests, such as thinning, climber-cutting, clearing fire-lines, fighting fires, road repairs and so forth. For his chief labour supply the Forest Officer was dependent on the local agriculture population, and he obtained it from neighbouring villages during the slack periods of agriculture. It was found that by concentrating part of this labour in the "forest village," to whom a certain area for agriculture and grazing purposes is allotted, the Forest Officer had a first call on the labour which the village could provide and was thus able to depend with certainty on what amounted to a permanent supply of labour at the period when he required it. He is thus to a great extent rendered independent of the great difficulties always experienced in the case of forest work in connection with imported labour. The institution of the forest village is capable of considerable extension both in Indian and in other parts of the Empire.[11]

The first attempt at establishing state control over forests in Madras Presidency was the introduction, under the East India Company, in 1807, of a state royalty over teak and other valuable timber in Malabar and Canara. There were many protests throughout the coastal areas of the presidency over what amounted to a government monopoly over timber in these regions; Sir Thomas Munro abolished this in 1822. The forests continued to be in charge of the district collectors, with no special arrangements for their use by the state. In 1847 the executive engineer of Malabar reported to the government that private contractors were denuding the forests in Malabar and Coimbatore. The government responded by appointing a special officer through the Public Works Department to extract teak for government use. It was not until Hugh Cleghorn was appointed conservator of forests of Madras in 1856 that there was any attempt to conserve forests rather than simply compete with other users for timber. Cleghorn recommended the abolition of shifting cultivation in 1860, but the government order to that effect was unsuccessful and unpopular, and was soon withdrawn (only to be reissued in 1882).[12]

Although environmental historians debate the content and scale of forest "conservation" versus "exploitation" before the advent of Crown rule, most are agreed that there was a definite change in the level of organization after 1858. The Indian Forest Service was formally established in 1865, in order to manage forest resources more efficiently in the light of the growing demand from the railways and other timber-hungry enterprises. Conservation, financial viability, and the meeting of village needs were the three main areas on the agenda of the Forest Service. The conflicting demands of these three imperatives led to much debate among foresters, revenue officials, and administrators. In 1878 and 1882, respectively, the Indian and Madras Forest Acts were passed, establishing considerable state control over forested lands, and strictly regulating local use of forests through the system of "reserved" and "protected" forests. These acts had immediate and far-reaching consequences for local villagers and tribes who had been accustomed to using forest resources under practically unrestricted circumstances.[13] In the last quarter of the nineteenth century, forest professionals surveyed, assessed, and developed working plans and communication facilities for forests all over the South Asian subcontinent, making forestry a commercially successful enterprise that produced reliable profits for the Indian colonial state.[14]

The Madras Forest Act (1882) was passed later than the India Forest Act (1878) because the revenue and forest officials in Madras Presidency were engaged in a debate over "customary use" (or the rights of local users) versus the importance of preserving forests. Although "village forests" were in fact established in many parts of the presidency, the Madras Act paid little more than lip service to the question of local need. It is sometimes argued that the Madras Act was formally more open to local needs, and that this was in fact the reason that this presidency saw less large-scale protest and conflict with the Forest Department through the second half of the nineteenth century. A detailed reading of the *Forest Proceedings* through the turn of the century shows, however, that there was in fact an almost daily struggle between the Forest Department and local users, culminating in the transferral of village forests to the Revenue Department in 1925 on grounds of non-profitability.

Nature as Property

One of the most eminent colonial foresters of the nineteenth century was B. H. Baden-Powell, well known for his extensive work on forest law and jurisprudence. At a forest conference a few years before the passing of the Indian Forest Act of 1878, Baden-Powell delivered an important lecture that explained the basis of the forest laws that would be required to bring about a scientific management of forests. He urged that it was imperative to "consider forests in a new light . . . as *pieces of property,* as 'estates' of a peculiar kind."[15] This constituted an important shift in the way nature was to be officially conceptualized.

The transition from the existing view of forests as an open resource to forests as property of the state was essential if the colonial state was to meet its timber needs. As we have seen in chapter 2, the Lockean principle of private property as constitutive of personhood was a powerful underlying naturalizing factor; that is, the classification of lands as private property was widely justified in terms of a natural progression from chaotic and uncivilized to ordered and modern society.

Baden-Powell stressed that the redefinition was "not only a matter of words. . . . It is an important and most practical conception. . . . For if you realise the idea of a forest estate to be cared for as a piece of property and protected by law, you will also acknowledge that a 'piece of property,' if it is to be either managed or protected, must be defined as to its limits, and all questions of right and obligation arising in those limits must be settled. If that is not done, the forest is still in a fluid, uncrystallised state; it hardly deserves to be called 'property,' and in consequence any real conservancy will be unattainable."[16]

Once its status as property was enshrined in law, he argued, forest land could be securely under the protection and exploitation of experts in the field: "law declares or recognises, that some persons have the *right* to appropriate things or become their owner; and other persons have the corresponding *obligation* to respect the right and to abstain from acts of interference, the law threatening some . . . punishment in case of disobedience. . . . Now, Forest officers are the managers and controllers of forest estates; and as such they enter into various relations with the public, and with individuals."[17]

Shortly after the conference at Simla, Baden-Powell published an article on "The Political Value of Forest Conservancy" in the *Indian Forester,* the journal of the Forest Department. He emphasized again that "Forest Conservancy starts from a basis of property." He stressed the importance of not giving in to the peoples' "clamour and complaint" over state control of forests, as this would weaken the state's image in their eyes and "the Government as a proprietor cannot safely allow what a private owner would not." Access to the forest must only be granted, he warned, "in such a way as to leave no doubt that it is regular recognition of right, by a power that is based on a respect for right." He went on to suggest, however, that the more fundamental right was that of the state to utilize the forests, while the local users would be granted occasional concessions. "When forest resources . . . are made available in an orderly and systematic manner, as distinct gifts of a power that might, as far as right is concerned, withhold them, . . . [the people will accept them as gifts] and so acquiesce gladly in the presence of a power, which, while it maintains its own in the interests of the whole country, gives what it is right to give, to the local requirements of the neighbourhood."[18]

It was both a moral and a political necessity for a state to assert its power this way, he argued. If a government was to be recognized as a strong power, it must show that it could exercise control over places distant from its ruling capital. "The existence of well-ordered and protected forest estates in the farthest and most out-of-the-way places demonstrates that completeness of control extending to the farthest verge of the State domain, which characterises a strong Government. In such distant localities there is often no other tangible proof of the Government power." He addressed an argument that was commonly made against such stringent measures to claim state proprietorship of the forests on the grounds that the resulting popular discontent was difficult to contain. Forest conservancy might be unpopular among "ignorant peasants," he agreed; however, the important thing was not to be popular, "but to be *just*."[19]

This counterargument introduces the moral charge with which foresters were regularly to invest their representations of nature. He chooses as analogies two of the most morally charged discourses of the colonial state, the struggles against tropical disease and "evil" social customs. "How should we act in a matter in which we *do* earnestly and practically believe?—e.g. in the case of suppression of 'satti' or in preventing cholera or small-pox? We do not there think of giving up the point; we say that the evil must be suppressed, come what may, and cast about to find the right way of doing it. That is exactly what I urge for forests. *There never was political danger in doing what is right and for the real benefit of the people, be they never so ignorant of the advantage.*"[20]

Richard Grove has described the moral urgency with which eighteenth-century colonial scientists had fought for state recognition of an "environmentalist" agenda. By the mid-nineteenth century they had won the interest of the state, and by the 1870s its support, although debates continued through the century over the relative merits of preserving nature and making quick profits. The Forest Department came under constant criticism from the Revenue Department for interfering with the practical job of state revenue collections, while the Forest Department in turn justified its slowness in making profits on the basis that it was performing important environmentalist tasks. By the end of the century, however, due to the persistent institutionalization of scientific management techniques and engineering methods, and the passing of Forest Laws in all parts of the country, the Forest Department was showing large profits.

Laboring for Progress: Disciplining the Madras Forest Tribes

In 1861, an inspection report of the Wynad forests complained of widespread evidence of slash-and-burn cultivation, but recorded that there were "acres and acres" of teak trees that promised potentially large revenues: "to clear out the Forests [of teak] thoroughly, it will take at least three years, so numerous are the trees."[21] The same report mentioned the presence of at least four hundred tribals, who "live entirely in the Forest. They are our only axemen, and without them it would be difficult to work a Forest."[22]

By 1866, there was a thriving teak plantation at Nilambur, in Wynad. Begun in 1844, it was one of the first systemic large-scale forest planting efforts of the colonial state. The *Administrative Report* argued that the hill tribes themselves would profit from access to markets, as this would free them from their dependence on middlemen, the Chetty traders who took the produce to markets. The benefit to the Forest Department would be twofold: again, these two factors were connected to local knowledge and a captive labor pool. "Forest products are in all cases collected by the Hill tribes, and the poorer classes residing at the foot of the hills or outskirts of the forests. . . . [They need] a regular market, such as the forest department might establish. A plan of this sort would tend greatly to the development of the resources of the forests, and many products, at present little known and never collected, might be brought into commercial use . . . and the fact of the Hill tribes and Jungle races being in a manner under the control of the Forest Department would be of great advantage to conservancy."[23]

By the 1930s, the "Report on the Aboriginal Tribes of Madras Presidency" was a regular feature of the annual *Administrative Reports*. The collector of Coimbatore reported that "touring cinemas" and improved communications had reduced isolation of aboriginal tribes; there were mission-run hospitals and schools in several tribal areas. From the Northern Circle, the forest conservator reported that "gradual, positive incentives to settled cultivation" were being made to the Koyas and Chenchus (two local tribes), while their slash-and-burn methods were being prohibited. The Forest Department had also begun schools for tribal children, and established health services that distributed free quinine. In order to encourage settled cultivation, the 1938 report records, "The Director of Agriculture has suggested the training of half a dozen Koya pupils in agricultural research stations for one year at Government cost so as to employ their services as agricultural demonstrators in the Agency to advise their fellow hillmen to take to improved methods of cultivation. This is also bound to be useful in inducing hillmen to take to settled cultivation."[24]

The report recommends encouraging immigration from the plains, commenting that this "means the opening up of the country and a pushing through of the road programme as rapidly as possible." For the Chenchus, a notoriously "unmanageable" tribe living in the Nallamalai forests in the northern part of Madras Presidency, twenty-three elementary schools had been established by 1938; there free meals, clothing, education, and quinine were provided to Chenchu children. In addition, free or subsidized cattle and land were granted to adults, who were hired by both Forest Department and private contractors (who were required to hire Chenchus for forest work in these regions).[25]

The Forest Department's settlement projects for the Chenchus are a particularly interesting example of their attempts to control tribals, as the Chenchus were perhaps the most resistant of all forest-dwelling groups to colonial civilizing efforts. The department went to great lengths to "domesticate" them, resorting at one point to classifying them under the Criminal Tribes Act; in 1925 they finally transferred them, along with all village forests, to the Revenue Department, complaining that civilizational projects were eating up their funds and not producing enough profits to justify the investment. (These same projects continued to be managed by the Forest Department, however, even after 1925, although they were paid for by the Revenue Department.) The measures taken by the Forest Department to organize the Chenchus into a labor pool illustrate the importance of tribal labor to the scientific forestry agenda.

In an effort to break into an apparently semifeudal system by which Hindu plains dwellers held the Chenchus in permanent debt and thus appropriated all their collected forest produce, the Forest Department paid off Chenchu debts to the traders and moneylenders. It would be difficult to argue, however, that this was a transitional step from feudal to wage labor, because the Chenchus were rarely paid actual wages by the Forest Department. Instead, the department would pay them for their labor with cloth, rice, and forest grazing or collecting concessions (that is, by waiving the grazing fees). The above-mentioned debts were actually paid off by deducting the sum from future wages, thus obliging the Chenchus to work for the department until this sum was paid off; their livestock was taken as security. The 1907–1908 Forest Administration report records:

> The Chenchus are a troublesome tribe with bad tradition and their reformation which is most desirable for the good condition of Nallmalai forests is no doubt an uphill work. But they are gradually being brought round through the indefatigable efforts of Mr Wood . . . and Mr Aswatham Naidu. Plantations have been opened in the different Chenchu centres and they are employed as firewatchers, in the collection of minor produce and in plantation works. To save them from the clutches of greedy *sowcars* [moneylenders] shops have been opened in the different Chenchu centres where grain, condiments and salt are sold at cost price. Cloths are also supplied to them departmentally and their cost recovered from the wages due to them for work.[26]

There was a permanent staff of forest subordinates stationed at each settlement; in addition, an assistant Chenchu officer would travel between all the settlements, having "little or no definite responsibility beyond trying the Chenchus who may be charged with the commission of various offences and punishing them according to law."[27]

In 1914, H. M. Hood of the Indian Civil Service was commissioned by the Revenue Department to do a six-month study on a "scheme for the control of the Chentsus of the Nallamalai Hills in the Kurnool district," and a "Special Chenchu Officer" was appointed in 1916, on the basis of his report. This officer (who in 1916 was Saunders, the assistant superintendent of police) was assigned the following duties:

> The Special Officer would be constantly moving from one Gudem [settlement] to another and would exercise close supervision by detailed and frequent inspection. . . . [He is] to supervise the conduct of the Chentsus and to devise means for the improvement of their condition. He should arrange suitable measures for their employment in communication with the District Forest Officer. He should exercise his influence in order to induce them to bring their children under instruction, to refrain from and prevent forest fires, to stop drinking and to secure the arrest of members of their community who have committed grave crimes.[28]

He was also to systematize agriculture; arrange loans for ploughs, bulls, and land; and stop Chenchus from leasing their (departmentally granted) land to settled cultivators. He was to establish a personal relationship of patronage with the tribals. "All rewards [for satisfactory forest work] should be awarded and distributed by the Special Officer and not by District Forest Officers. The distribution should be made by him personally, and on a consideration of whole conduct of the gudem concerned, and not merely their conduct in connection with the Forest Department and its work; their conduct in respect of the forests would of course be an important factor in deciding the . . . awards."[29]

By 1928 the department argued that the Chenchus had many sources of income. "If in spite of all these sources of income, some of the Chenchus still commit occasional crimes, the remedy appears to lie in introducing effective methods of settling them down to the practice of agriculture, with the help of trained agricultural demonstrators." The assistant Chenchu officer, the conservator ordered, "must do substantial missionary work in the cause of agriculture." Reviewing departmental Chenchu schools, he reminded officers of the objects of school. "It is a first step in reclaiming a tribe of criminal tendencies and making its members useful members of society and less of a menace to the public welfare."[30]

The claim that tribals were a menace to the forest is, on the surface, a contradictory one, since the early foresters themselves had often learned from tribal knowledge of trees and forest products. However, we can understand this contradiction as functional rather than simply incoherent. Through the end of the nineteenth century, the Forest Department was trying to establish as superior a system of production that required the creation of an orderly and disciplined labor force, and to institute an equally radical transformation in the forest itself, involving the extension of growth in commercially valuable species and the elimination of commercially useless species in order to achieve that cherished silvicultural goal, the "normal" forest.[31]

The Koyas objected to the practice of clear-felling stands of large trees, on the basis that it conflicted with their notions of what was sustainable. When they practiced forest cultivation, they felled only the smaller trees, and ones that they could use as firewood and building material. When the Chenchus wanted to collect large quantities of the ippa flower, they fired the undergrowth in order to eliminate forest growth that was not valuable so as to allow easier access to the desired forest growth, the ippa flower. Although the Forest Department followed the same principle when clear felling a region in order to plant or isolate a more valued species, they regarded the Chenchu habit as evil and destructive of nature, while representing their own methods as nature-friendly.

The following note on silviculture illustrates what foresters understood by the scientific improvement of a forest. Cutting or burning were appropriate as long as they were done for scientific and progressive purposes: "The object in view must not be lost sight of. It is to improve the forest and to bring it gradually closer to normality. . . . Felling by daily wage labour, to be controlled strictly, will involve the cutting or girdling of valueless growth interfering with valuable trees, or the regrowth expected from valuable species cut back."[32] Clearly, then, the practices of cutting trees and burning undergrowth per se cannot be understood as inherently good or bad for nature. Rather, the historian must follow these practices as they are refracted through an ideological system in which the rhetoric of protecting nature is of a piece with a system of production, and a corresponding attitude to labor management, in the context of the increasing social and political importance of scientific and technical knowledge.

The Chenchus were the most refractory of the Forest Department's wards; however, the measures taken here were identical to the systems followed in all tribal areas. In June 1921, as a result of an ongoing discussion between forest conservators all over the presidency, a conference was called at Ootacamund between the chief conservator of forests (S. Cox), the protector of the depressed classes (T. E. Moir), and the director of public instruction (R. Littlehailes). They resolved that there was a "need for the institution of special forest preparatory schools for the education of the jungle tribes, in cases where the Forest Department is dependent on such tribes for labour." Free midday meals were to be given. Although several forest circles already ran elementary schools, the resolution stressed that it was important to continue training beyond primary school, so as to prevent "losing 13 to 14 year olds," who would finish school before they were old enough to be hired by the department. Therefore it was recommended that after age thirteen, the children receive "vocational" training that would prepare them to be hired as forest guards.[33]

One report that had led up to this conference was written shortly after a local famine in May 1919, when attendance in schools declined as children spent all day in search of roots and other food from the forest. The district forest officer wrote:

> These Sholagas are on the whole a people whom we desire to retain and improve. I do not approve of withdrawing from a school once started. . . . It will have a bad moral effect . . . being semi-savages they do not know what will benefit them, or if they know they still find it difficult to make present sacrifices themselves for the benefit of their children who should make good foresters if educated very little. It is very difficult for them to manage unaided and so I think we should give them one square meal a day all the year round and get things to move quickly. Then we can have real control and enforce discipline. Better spend a little more and make the affair a success than drag on with a few hungry children. After all, the improved health means better men for us in the end. And here not only is it desirable from an ethical academic point of view to help the Sholagas on, but we have to remember that if we do not get them to lean on us we shall lose them i.e. they will follow their pals and become timber-thieving Sivachari slaves—a poisonous semi-forest tribe to have to reckon with—or else only retain the looting scum.[34]

Here a discourse of moral progress from savage to civilized piggybacks on a quite explicit argument for the economic interests of the Forest Department. Control over a labor force and healthy, disciplined foresters are the desired ends of the department; these are cast in terms of a positively valued ethic of forest work (as opposed to natives cutting timber for their own profit) and schooling as an intrinsically moral, civilizing process.

The belief that disciplined forest work was morally superior to tribal livelihoods led some foresters to suggest that Christian missions, funded by the Education Department, be brought into tribal settlements in order to civilize them and train them in modern work routines. This was a particularly significant move, since colonial state policy at the time was dominated by the Utilitarians, who strongly discouraged proselytizing. Conservator A. F. Minchin wrote: "If financial assistance is to be given by the Educational Dept and if no political objection is to be raised to a little religious instruction, it seems likely that the Director of Public Instruction and the Mission authorities would arrive at a satisfactory working arrangement."[35] The chief conservator of forests in Madras, S. Cox, resolved that since the "financial and practical running" of Sholaga schools was difficult for the Forest Department to sustain, "I would propose to hand over the management, subject to [general control by district forest officer], to one of the missions at Satyamangalam. The Roman Catholic mission has a considerable settlement a few miles away. . . . I have very little doubt that [they] would avail themselves of the opportunity. . . . The Kavalur school, I suggest, should be offered to the Roman Catholic Mission and the Uginian School to the Protestant Mission."[36]

Although the Utilitarians took strong positions in favor of science and against religion, the everyday workings of science and Christianity were in reality not radically opposed to one another. Christian missions' work was imbricated with scientific and technological modernity in numerous ways (some of which I explore in chapter 6). The Basel Mission Society, the most active mission in Malabar, had a special division, known as the Cooly Mission, which would preach to plantation workers on a regular basis, including within its gospel instructions about work habits, discipline, and sanitary living.

By the 1920s, plantations were established all over the Malabar and Wynad area, and the Forest Department had to compete with private employers for labor. Although planters imported labor from the plains, they also employed a large number of local tribals. In 1929 the chief conservator noted that "In general, it is believed that, on equal pay, the Kurumbas prefer to work in the coffee estates rather than in the forests."[37] He urged that every attempt should be made to settle the tribes in the forests so as to "stabilise and enhance" the local labor population. "Every inducement must be offered, and free of all restrictions if necessary, to populate every swamp in reserved forests which is large enough to support a family. If the grant of *takhal* [slash-and-burn cultivation] becomes essential in order to retain Kurumbas, it may be given in bamboo forest or in the regeneration area. All such land, however, must receive some treatment, even if it be only the broadcasting of teak seed, to lead towards an eventual re-establishment of a tree crop. The order that 'labour must be imported and it must be treated precisely as planters treat theirs' has been made at intervals since 1893."[38]

He recommended, in addition, calling for inspections by a malariologist with a view to instituting antimalarial measures, forming more schools, and giving grants for purchasing ploughing bullocks. He called for district forest officers to develop a personal relationship with the tribes, as the "labour situation in the Wynaad is still very unsettled . . . success must depend on the local experience and personality of the District Forest Officer and of rangers." The report emphasized that the "valuable timber must be exploited departmentally; the Kurumbas must fell all other growth"; and that this should be closely supervised; "elephants should not be lent to the Kurumbas to assist in clearing."[39]

The Forest Department took care not to disrupt existing stratified, caste-based relations of production when they could be used to their own ends. So, for example, they would obtain Paniyars for forest work by making agreements with the Chetties, to whom the Paniyars were bonded laborers.

> The chief relationship between the Forest Department and the Chetties lies in the supply of Paniyar labour. The Paniyars are virtually the slaves of the Chetties: they own neither land nor cattle. On the whole, they seem content with their lot—or resigned to it—and they appear to have little ambition towards social betterment. Their normal work is in the cultivation of the Chetti paddy fields; but they can sometimes be obtained for work in the forest; this being largely contingent upon the relations, cordial or otherwise, subsisting between their Chetti masters and the forest subordinates.[40]

The Forest Department also closely investigated the organization and segmentation of tribal groups; these investigations were regularly reported in the *Indian Forester*, in styles closely resembling ethnographic studies of the time. These were, however, not mere academic contributions to ethnography but were directly useful in categorizing and predicting labor patterns. So, for example, the segmentation of the Kurumbas into three subtribes was noted as follows.

> The *Mullu* Kurumbas cultivate wet lands [for which rent was waived since 1915]. . . . They want little from the forest other than land to cultivate and the concessions allowed to jungle tribes resident in the forest. They work for the dept, more or less as required. The *Bet* Kurumbas, who are the local axemen *par excellence,* favour *takhal,* or shifting cultivation in virgin forest. . . . They do much work for neighbouring estates, but the forest residents depend very largely

> upon the department for their livelihood. . . . The *Jen* Kurumbas . . . are the least civilized. . . . [They do some takhal cultivation, but mostly collect honey, roots, and fruits] . . . local labour is not always easy to obtain for any length of time . . . the *Jen* Kurumbar will only gradually be habituated to regular work.[41]

The different groups would not cultivate in each other's forest areas; the Forest Department tried to alter this reluctance, but noted that there were signs of discomfort with this. "There are signs that this reluctance is being overcome since Malabar Kurumbas have been persuaded to grow ragi in the Bhoothakal regeneration area. It is possible that an outbreak of smallpox at the time of the writing may be considered retribution for this departure from tradition."[42] Such details—the attention to subdivisions in tribal groupings, their lifestyles and customs, and the myths or cultural beliefs associated with particular activities—are usually considered the province of ethnographers; however, these were a common part of regular forest reports.

Similarly, a report several years later, from a district forest officer in Coimbatore, shows careful attention to the intergroup dynamics of Nilgiri tribes. Having observed the local tribal interactions over a number of years, he concludes that the Badagas and the Sholagas are interdependent tribes, and that a successful settlement operation should provide crop lands for both. The Badagas, he suggests, would be of use to the Forest Department. Being cattle breeders, they "would be of considerable assistance in problems of grazing administration, and in experimenting on the reclamation of swamp lands to effective pasture." He describes his argument as follows:

> I have heard the lease of land to the Badagas criticised as they are little more than professional cattle breeders and do not contribute much labour to the department. On careful enquiry I am much convinced that their presence at Geddesal is necessary for the well being of the Sholagas on whom we depend for a good deal of labour. The Badaga pen a total of some 300 head of cattle and graze the surroundings of the settlement, thus reducing the fire hazard, and contributing a good deal of manure to fertilize the settlement lands. Their cattle have converted what used to be a veritable swamp carrying a heavy crop of swamp grass into a decent grazing ground in which an increasing proportion of good palatable grasses are appearing. They lend the Sholagas plough bulls at reasonable rates. The swamp used to be a playground for elephants which did a great deal

> of damage to the crops, whereas now elephants are comparatively rare. Were the Badagas to be turned out, the position would rapidly revert to that of ten years ago. To keep the Badagas permanently settled it is necessary that they should have land, a few acres of which they will cultivate, while on the rest they can pen their cattle and move their pens frequently, thus avoiding the creation of unhygienic penning conditions.[43]

The Sholaga settlement at Geddesal had been started several years before by missionaries, who introduced them to cultivation; when the missionaries left, the Badagas migrated into the settlement and began grazing their cattle.

Here we can see the intersection of tribal patterns with missionary and Forest Department interventions, resulting in a system that supplied the needs of the department, and brought tribal groups into configurations of production and interdependence that were different from precolonial ones in that their production now was oriented toward state interests. It should be noted, however, that this picture is quite different from the model that several environmental historians have drawn, in which colonialism is seen as a watershed that radically divides tradition from scientific modernity.[44] The advent of colonial forestry, it has been suggested, destroyed existing systems of resource use and replaced them with crude coercive regimes of exploitation, in which nature and the natives were summarily destroyed. The everyday functioning of forest management, as revealed in *Forest Department Proceedings* throughout the colonial period, however, suggests that the close attention paid by scientific forestry to local social dynamics, and the everyday, ad hoc adaptations in both the functioning of "modern" production and in preexisting forms of local use are important aspects of the specific forms of modernity that were being ushered in during the late nineteenth and the early twentieth centuries. In other words, modernity is not as pure of premodern elements as is suggested by historiographical models that attempt to separate the spheres of science, culture, and ideology.

To contest the model that portrays colonial environmental change as alien and destructive is not, however, to be an apologist for colonial science. It is, rather, a plea for historical rigor, in that it complicates our models of change to understand that tradition and modernity were not easily separable systems of knowledge; that preexisting and newly developed forms of knowledge entered into a complex, mutually constitutive relationship; and that the resulting discourses and practices of modernity were significantly influenced by and dependent upon the multiple intersecting, stratified, and nonstatic forms of indigenous production they encountered every day. In attempting to improve our understanding of scientific modernity, we can more clearly see some of the origins of what today appear to be self-contradictions in postcolonial modernities.

While Forest Department methods were elaborate and multilayered, the tribes were by no means passive victims of evil scientific manipulations. There is, in the *Forest Proceedings* themselves, a rich record of disobedience toward state efforts to discipline tribal populations.

Resistance

From the beginning of European activity in Madras forests, there is a record of indigenous efforts to retain control of these forests.[45] The record of tribal resistance must of necessity be read through official forestry reports, as tribal groups rarely had the access to laws and literacy that would have been required in order to submit petitions and thus leave records of their own positions.[46]

Around 1921, a new section appeared in the annual *Forest Administration Reports*, headed "Disturbance by Non-cooperators." The 1921 *Report* records trouble in many parts of Madras Presidency; the police arrested and sentenced several tribals; and an Indian member of the Legislative Council was called in to give propaganda speeches in support of the Forest Department. The report recounts:

> Cattle being rescued [from impoundings], cairns demolished, wholesale raids for green leaf manure, . . . subordinates assaulted. . . . The general lawlessness induced by the "Non-cooperation" lecturers also led to an increase in forest fires, and to other offences. The receipts given for collections made for Congress purposes were alleged by the ignorant payers to cover the right to graze in the reserve in one neighbourhood. . . . In Tiruvannamalai, three non-cooperating villagers were dealt with by transferring the grazing privileges conferred on them to their neighbours.[47]

In Malabar, several forest guards and peons were killed in connection with the Mapilla rebellion, which was primarily an agrarian dispute, but one which also inspired extensive violence in the forest areas of Malabar.[48]

> The outstanding event of the year as far as this circle is concerned was the Mapilla rebellion. . . . [Several people were killed,] though

> the DFO and the rest of the establishment managed to escape with their lives, they experienced much hardship and lost all their belongings. . . . Practically all the forest buildings in Nilambur division were looted and burnt and 75 of the 85 buffaloes were killed by the rebels. . . . Range offices burnt, quarters looted, all records were burnt and work was entirely dislocated for about 6 months and it was impossible to float any timber down to the coast; hence very little revenue was realised during the year.[49]

The 1923 *Administrative Report* records an increase in illegal timber felling after the rebellion. At the same time, in the northern forests of the presidency the hill village system, by now well established, was facing resistance. The chief conservator of forests noted in 1921 that many Malayalees withdrew from the hill village system, "but realized that they lose by doing so."[50] In rejecting the paternal protection of the Forest Department, the Malayalees found themselves having to pay Rs. 1,400 per annum in fees for grazing and collecting forest produce, activities over which the Forest Department had a monopoly and for access to which they charged high fees. Malayalees were reported, in this period, as stealing sandalwood, and using "the knives which they were allowed to carry to cut creepers and climbers, for lopping the tops of branches and bamboos to feed their cattle . . . and warnings and fines have not produced any marked effect." Further, "Assaults on forest subordinates were rather a feature of the period in Anantapur district. . . . The Forest staff found itself more or less helpless in the face of organized unlawfulness which had been fomented by the preaching of the non-co-operators."[51]

This report also has a long meditation on the cunning behavior of the Khonds.

> The relationship of the Khonds to the Forest Department would appear to resemble that of thoroughly spoilt children. . . . Like other spoilt children, the Khond is in no way averse to doing what he knows he should not do, provided he has a reasonably sporting chance of escaping detection and avoiding punishment. . . . He knows, in fact, that he may even gain a certain amount of kudos over the business if he plays his cards skillfully. . . . During the clearing of the fire lines he works willingly and creates a very good impression and earns a certain amount of money. Later on, probably about noon on a day in the hottest weather, he arranges for fires to be lighted inside the

> fire lines in 5 or 6 places almost simultaneously. By evening the local forest guards will have seen one of these fires and will arrive at the village demanding help to put it out. The Khond, with his tongue in his cheek, will turn out his village and take a number of men to help in extinguishing the fire. In the meantime the other fires will have got beyond all control and in due course the District Forest Officer will receive a report saying that the Khonds rendered every assistance and were of great help to the department but that unfortunately, say, six thousand acres of forest were burnt. The above is the impression which the present writer has formed of the Khond after a brief acquaintance with him, covering some 3 months of the worst fire season that he has ever experienced. . . . The writer of this report . . . would find it difficult to believe that the Khonds of Ganjam can compete in devilry with the Chenchus of West Kurnool. . . . The Koyas and Reddis of the Godavari divisions continued to be serviceable as usual . . . only trouble is their drunkenness.[52]

Also in 1921, a missionary traveling in Salem wrote to her mother: "The ignorant have been stirred up by the [Congress] agitators to believe that Gandhi is King now, and that the British rule is at an end—the results being that the villagers have been trespassing in the reserved forests and taking leaves and branches for firewood ad lib, and some are now in Salem gaol on that account."[53]

In 1924 the *Proceedings* of the chief conservator of forests announced that the hill forest village system in the Yelagiris and Tirupattur Javadis was a failure "due to (1) outside teaching (2) the well-to-do condition of the Malayalees (3) existence of large tracts of unreserves; but mainly due to the Malayalee character."[54] The phrase about "outside teaching" refers to the influence of the Non-Cooperation activists who traveled through the forests and villages spreading the word about the national movement. The reference to "the Malayalee character" indicates that even in the presence of explicit political explanations for resistance, foresters continued to have recourse to an essentializing discourse of inherent tribal lethargy.

There are some instances of protest petitions being submitted to the Forest Department. In September 1935, the South Kanara Catholic Association sent a memorandum to the governor of Madras. They wrote, "The members of our community are keenly affected by the withdrawal of concessions granted to villagers in Reserve Forests in this district." They cite forest policy, quoting the Government of India Resolution of 1894, included in an appendix of the Forest Code: "It is laid down that every reasonable facility should be afforded to the people concerned for the full and easy satisfaction of their needs for fuel, manure, fodder, etc., and that considerations of Forest-income are to be subordinated to that consideration." The secretary to the governor commented on the report that this was an "old, old story: already fully noted on for H. E." Attached to the memo was an "office note": "The complaints . . . are not peculiar to the Roman Catholics of S. Kanara alone. They apply to all the ryots of S. Kanara and relate to the withdrawal of certain concessions granted by the Chief Conservator of Forests in 1922 and which have since been withdrawn in respect of 17 reserves owing to abuse."[55]

It was in 1924–1925 that the Forest Department formally transferred responsibility for the hill villages, or ryot forests, to the Revenue Department, so that "trained talent" could concentrate on developing "valuable property" and profits. The 1924 review of Forest Department work noted that "In 1925–6 we expect to begin to reap some of the returns from organizing the business side of the Forest Dept along business and engineering lines." Conservator C. S. Martin predicted: "Within 10 years I believe we can attain a surplus of 40 lakhs."[56]

After 1919, in fact, there was a significant shift in the management goals of the Forest Department. The foresters had found themselves unable to meet the increased wartime demand for timber, and the postwar period found them resolving to update their science to fit the new age. The new science of the time, of course, was applied engineering. The department sought out expert engineering advice on grading and testing of timber and on transport and extraction techniques. C. S. Martin's 1925 definition of engineering is instructive: "From an economical and practical viewpoint, engineering may be defined as the science of expending a rupee so as to bring the greatest returns. Logging Engineering, then, makes the forest rupee bring the greatest returns."[57]

By now the easily accessible timber supplies had been exhausted, and the foresters had to go further up the mountains in order to find fresh supplies. This brought new topographical difficulties to be surmounted. New mechanical logging methods were required, making logging a highly technical exercise. Forest engineers also needed to understand the principles of silviculture and scientific forest management, as these older sciences were by no means superseded by the new developments in technology. "The most efficient harvesting of a forest crop requires also a knowledge of trees, their characteristics, habits, state of health, maturity, etc. . . . [T]herefore a logging engineer needs to study 1. forestry, 2. physical science—civil, mechanical, electrical and hydraulic engineering, 3. technical logging methods, 4. cost accounts . . . so as to ensure sustained efficient yields."[58]

Changes in technical knowledge were thus incorporated into the existing commercial framework of operation of the Forest Department; the scientific forestry of eighteenth-century Germany was in the early twentieth century modernized through the infusion of contemporary engineering techniques.

The tighter postwar style of management was fairly successful in containing the threat of increased tribal resistance to forest policy. To a large extent, systems of domestication and control of the tribes had already been laid down, the major part of the organizational work for the hill village system having been done in the second half of the nineteenth century. Even the large-scale disobedience practiced during the 1920s and 1930s failed to dislodge these systems of control; in fact, we can see them embed themselves more rigidly through even more modern methods of disciplining, such as apparently noncoercive propaganda methods. For instance, the 1924 review of forests reports that Durbar Day had been celebrated in the Tiruvannamalai forest division. "Durbar Day . . . was attended by the surrounding village panchayatdars and forest villagers, and officials and citizens of the taluk. The function was a success and went a great way to demonstrate the advantage of the protection of forest and excellent system of cooperation amongst the ryot population and of the forest village system by which good protection of reserved forests was being achieved."[59] Even as protest was asserting itself, systems of control and propaganda were becoming more sophisticated, and presenting themselves in the form of popular programs.[60]

The same report also makes a positive assessment of the tribal labor question, reporting that on the whole there are numerous efficiently functioning forest panchayats, due to the forest education propaganda. Unsurprisingly, the rhetorical assertion of a peaceable kingdom of happy tribals cooperating with scientific foresters is a clue to a glossed-over contradiction of this state of affairs. Indeed, reports as late as the 1930s indicate that forest protest had by no means died down. In Coondapur in 1933, three villages had wilfully dissolved their forest panchayats, were "clamouring for added concessions," and had "laid claim to what amounts to a complete ownership of the forest."[61] All over Madras Presidency, tribes continued to resist the restrictions on their activities. In the same year, "Savaras of the Parlakimedi Malais are not yet reconciled to the curtailment of the area formerly freely enjoyed by them for podu, by the reservation of certain hill-blocks under section 26 of the Forest Act. Some Savaras who ventured to clear a block of land were convicted. As the Savaras are refractory and their population is said to be increasing, the problem of providing an alternative means of livelihood for them is difficult."[62] And in Cuddapah in 1937, The villagers obstructed the forest subordinates and forcibly removed the goats [that were browsing illicitly] before they could be impounded."[63]

Globalization and Modernity

The conclusion of Chief Conservator C. S. Martin's 1924 report calls attention to the increasing economic importance of updating the technologies of forest management. "The economic development of tropical forests is being forced all over the world by the exhaustion of . . . timber. In the Philippines, Formosa, Borneo, Java, Mexico, Brazil, and Central America, new capital is constantly going into forest enterprises, and it certainly behoves Madras with its large unused resources to keep pace."[64]

By the end of the first quarter of the twentieth century, scientific forestry had entered a new phase: control over native labor had been more or less consolidated, and protest, though frequent, was contained; meanwhile, the German science of silviculture had been complemented by the more modern techniques of engineering, and the economic development of forests was becoming more and more central to the global commercial requirements of colonial modernity.

One might plot the development of forests' "modernity" according to the increasing technological content and global reach of forest activities. But such a plot does not reveal a corresponding liberalization in labor laws, or a diffusion of power to local users of the forests. Rather, modernization was accompanied by the entrenchment of coercive, propagandistic regimes of labor and the continuation of a moralizing rhetoric of progress from savagery to civilization through the discipline of work.

Far from being an anomaly, or a failure of colonial modernization attempts, as some historians of technology would suggest, this so-called premodern aspect of the relations of production was a structural requirement for the corresponding global expansion of the colonial economy.[65] Does this mean, one might ask, that the colonies failed to enter a truly scientific modernity in the early twentieth century, when European countries were in fact bringing their social systems into line with the modern technologies of production and management? It would follow from such a hypothesis that the premodern elements of colonial societies would be expected to wither away with the continuing development of scientific systems. Such an assumption is, indeed, commonly made in popular accounts of postcolonial societies.[66]

These so-called premodern elements persist alongside an explicit state policy of embracing large scientific and technological development projects, giving top priority to science education, and imparting an almost sacred cultural status to science. Although the modernizers would assert that this persistence is evidence of the state's failure to erase completely all elements of a prerational world, the nostalgic environmentalists see these remnants as the only hope for a return to a pastoral pretechnological utopia.

But if we start with a more historically accurate account of the constitution of colonial modernities, we can no longer hold on to the dichotomous representations of modernity that these accounts implicitly assume. "Premodern" elements were functional and essential aspects of the constitution of colonial modernity. They were not just manifestations of an unfortunate but accidental persistence of older forms due to native reluctance to give up old ways or insufficient administrative attention to science education on the part of the colonizers. Nor were they innate antimodern aspects of tribal identity, inherently inimical to science and inaccessible to analysis. Rather, the rigidification and retention of so called premodern elements were an essential part of the functioning of the colonial economy. Through the interweaving of political, economic, cultural, and technological discourses and practices, we can see the construction of nature and natives as resources for colonial production. This interweaving of elements produces what I will call a "mixed" modernity. In the next chapter I explore the interdependent constructions of nature, natives, and modernity in the context of coffee and tea plantations in the southwestern hills of India.

Chapter 4

Plantations

Recruiting and Managing Plantation Labor

"Are we not safe even in our village?" asked Karupan.

"No, the white doraies [bosses] are all powerful. You see they rule the country. So the police have to serve their interests. When once a worker enters an estate, it is very difficult to get out of the clutches of the maistries [plantation contractors and/or foremen] and their masters, the white doraies. The advance which a man gets in the beginning will be his shackles. Mostly the coolies [plantation workers] will never be able to repay this amount at the end of the first year. The maistries will see to it that you are not able to clear the debt and thereby your services are secured for another year. As you know, the wages are paid only once in a year, during settlement. The advance which you receive in the beginning, the travelling expenses, the cost of the kambly [blanket], the cost of the weekly rations and selvoo cash [weekly allowance] . . . will be deducted from your wages. You know we have the hill fever [malaria] here. Most of the coolies will suffer two or three attacks of the fever. If you are ill and unable to work, the head field writer may order a part of the weekly ration to be given to you. . . . At the end of the year they will produce an account which will surprise you. If you question the correctness of this account, you will be called an ungrateful dog and sometimes blows will be the result. So, you have no alternative except to borrow or starve. If you are a single person your wages at the end of the year will never be enough to clear your debts and you are chained to the estate for another year. If you have your wife also working and if both of you are fortunate enough not to fall ill you may be able to clear the debt at the end of the first year. But this is rare as every one will surely fall ill many times. But if in a family there are two or three children who also work, they may earn enough at the end of the first year to clear their debts and go home. That is why you will see even the small children working. . . . The white doraies rule the country and you cannot escape from them."[1]

South Indian plantations recruited labor through a system similar to that used with the Chenchus by the Forest Department, but more coercive. Labor contractors (known as *kanganies* or *maistries,* usually Hindus of higher caste than the laborers) would offer cash advances to landless laborers from the plains in order to induce them to move to the hills. Labor historian Ranajit Das Gupta notes that this often simply meant that debt was transferred from the hereditary landholder or moneylender to the planter and contractor.[2] Workers were not permitted to leave until their debt was paid off, which in many cases took years of work on the plantation—a situation economic historians characterize as "debt bondage."

Although plantations were seen by the state and by foresters as beacons of civilizational progress, their success was dependent on the incorporation of modified "pre-modern" elements such as the transfer to planters of caste-related indebtedness, and nonmodern patterns of work, including extensive manual labor, long hours, and coercive supervision of an essentially captive labor force. Plantations marked a shift in agriculture from subsistence to capitalist production. Most stage-based economic models of modernization assume that the shift from subsistence farming to capitalist peasant relations is accompanied by a shift from feudal to mobile, free wage labor.[3] As Das Gupta points out, however, this did not happen. "An interplay of indebtedness, caste hierarchy, personal dependency relations, and *sahib-power* backed by the state, helped to keep workers under permanent control and in a state of perpetual fear, and significantly to restrict the operation of the labour market."[4]

The perpetuation of preindustrial modes of stratification leads Das Gupta to see plantations in colonial India as characterized by "unfree labour" and "a highly authoritarian structure of management of labour." This system of managing labor was further facilitated by what Das Gupta refers to as the "political geography" of south Indian plantations. Plantations were invariably situated in "virtually unexplored and/or very sparsely populated hill and forest tracts." In 1901 the population per square mile was 116 in the Nilgiris (which had mainly tea plantations), 114 in Coorg (a coffee-producing region), and less than 100 in the Wynad highlands in Malabar (which had both tea and coffee plantations).[5] This isolation made possible a disciplining of labor that was almost absolute.

Much of the official record of this period presents us with the planters' point of view, in which sustained production for the global market was the primary aim. P. H. Daniel's novelized account of labor conditions in south

India, *Red Tea*, describes living conditions on plantations.[6] In this novel, Daniel details the process by which advances are given and debt bondage occurs, the inadequate housing and poor health facilities, and the near impossibility of escape from this predicament. One of the characters explains to a newly arrived laborer why he cannot leave.

> "I have made up my mind to go and we will walk the distance and beg for food on the way," defiantly spoke Karupan. . . .
>
> "Even if you sneak out of the estate, you will never reach Pollachi," said Muthiah. "You may remember a place called Oomayandi Mudaku, while coming up from Pollachi. And you may remember seeing a building by the road side in this place which is occupied by a doctor and a peon. Well, they are stationed there by the white sahibs to prevent coolies from running away. On one side of the road is a steep hill which no one can climb, and on the other is a deep ravine. So you cannot pass the place except by the road. And you will not be allowed to do so, unless you have a pass from the estate. If you do not have a pass, you will be sent back to the estate under escort and then it will be hell for you. There have been cases where such run away coolies have been beaten to death by the maistries. So, remove from your mind, once and for all, any idea of leaving the estate."[7]

The plantation economy had a huge impact on tribal groups who had freely used the forests and hills that were now being converted into ordered productive landscapes. Although a major part of plantation labor was imported from the plains, the early planters were very dependent on tribal labor for clearing the jungle. Tribals continued to be hired through the turn of the century, as we can see from the forest records, in which foresters record competition with the planters for tribal labor. In 1845, when the Ouchterlony Valley Estates began in the Nilgiris, peasants from Coimbatore, Salem, South Canara, and Malabar were brought up to the hills, but would almost immediately "bolt away." Recalling this in 1896, an article in *Planting Opinion* recounted: "But in the course of time, the Cooroombers in the Devlala district were persuaded to come to the valley—and these men were of great help, because they are excellent axemen. The rate paid for labour then was very low, 10 annas men and 1 anna 4 pies women. Cooroombers were regularly employed at felling. The wages began to rise from 1858. . . . There is no scarcity now, as there are any number of 'locals' to be got, namely Cooroombers, Panniers, Badagas, and Chetties, who are the aborigines of the country."[8]

In 1862–1863 the first group of planters started the South East Wynad Planters Association, which successfully petitioned the government for good roads, post offices, and hospitals. The government, happy to assist in the opening up of the undeveloped hill areas, also built a motorable road from Ootacamund to the Malabar coast, and annexed South East Wynad District to the Nilgiris District in order to make administration easier. The early planters made many references to the useful work of the west coast Roman Catholic missions, which built churches and made regular visits to plantation areas.

In 1920, the United Planters Association of South India (UPASI) issued a statement on labor that called for the collection of ethnographic information on the natives so as to facilitate labor recruitment. It stated that the "Committee believe that too little is known of the customs, religions, castes, and prejudices of Indians, and that our knowledge would be improved by the appointment of an Agent, to be styled . . . the 'Indian Political Agent,' a Civil Servant, who would reside in India and whose sole duties would be to study India and Indians, for the benefit of employers of Indian labour."[9] Knowledge about the customs of the laboring castes was invariably rendered in terms of supposedly inherent tendencies, and in terms of rigid caste hierarchies and feudal structures from which modern forms of employment might rescue them. Introducing natives to the habits of industriousness was the formal legitimation for most labor practices of this period.

In his memoirs, Somerset Playne, a planter, recounts the history of the Ouchterlony Estates (the earliest Nilgiri plantations). He describes the "Bet and Jain Kurumbers" as "wild tribes" who were enslaved by the Chetties. The early planters saw themselves as rescuing such tribes from slavery based on caste hierarchies. Playne recounts the following.

> When the late Mr Henry Wapshare, the father of the present manager [of the Ouchterlony Valley Estates] . . . was felling the jungle for cultivating coffee . . . he suddenly came upon a gang of Jain Kurumbers frightened and huddled together. He surrounded them with the Canarese coolies who were felling the jungle at the time, and took them to his bungalow, shut them up in a godown, and fed and clothed them. When captured they were almost perfectly nude. After 3 or 4 days he released them, by which time they had realized instinctively the nature of the late Mr Henry Wapshare's object; and this fact, coupled with his kind and generous treatment, so impressed these men of the jungle that ever afterwards they used to flock to his bungalow with gifts of honey and other produce. This was in the

> early sixties. To this very day may be seen descendants of this particular tribe working industriously on the Naduvattam and Kelly estates.[10]

This narrative assumes the uplifting character of work, and represents the planter as a kind, paternal provider of this opportunity for upliftment and advancement. Although tribal laborers made up only a small proportion of the total plantation labor force, they were considered valuable resources, especially in the early- to mid-nineteenth century, when new plantations were opening up the hill areas. The representation of work as the route to civilization, which we have already seen in the case of forest labor, was not, however, restricted to tribal labor. As we will see, this rhetoric persisted well into the twentieth century, and was applied to Tamil labor from the plains as readily as to tribal labor from the hills.

It is interesting to contrast this situation with the changing attitudes toward labor in Europe during this period. Historian of science Anson Rabinbach, in his intellectual history of labor and technology, *The Human Motor*, suggests that in the early and mid-nineteenth century, discourses around labor were very much part of a moral and racialized discourse, in which both white working classes and nonwhite races were considered dangerous and dissolute by virtue of an incapacity for work, thus needing direct surveillance and moral instruction in order to advance along a civilizational scale of industriousness.[11] By the end of the century, he asserts, the morally inflected discourse of idleness had given way to a scientistic discourse of energy, directly influenced by the new developments in thermodynamics and the physics of motors. "Toward the end of the nineteenth century, . . . the ideal of a worker guided by either spiritual authority or direct control and surveillance gave way to the image of a body directed by its own internal mechanisms, a human 'motor.'"[12]

Rabinbach sees this transition from a moralistic to a positivistic understanding of labor as the very basis of social modernity. The doctrine of energy (beginning with the physicist Herman von Helmholtz), he argues, "contributed to a decisive break" with the premodern moral discourses of idleness. Scientists and reformers in the 1890s sought, through objective science, to "end [the] cacophony of moralising claims" about labor in which "ethics, science and politics were entwined. . . . As the condemnation of idleness was appropriate to a society with its moorings in religious conviction, the calculus of energy and fatigue was syncretic with a more scientific age."[13]

In the colonies, however, a moralizing discourse persisted well into the twentieth century. A major issue for the planters was how to control the coolies' drinking (recall that this was also an important issue for the foresters in connection with tribal groups).

> Should the Tamil cooly have perfect freedom as to the amount of drink he may consume provided its cost remains within his purchasing power? We suppose that, if we look at the question purely from the greatly abused principle of the freedom of the subject, he should. . . . A European is allowed to do so. Why then should not the estate labourer? To hang on, without reason, to what is regarded as the principle of equality for all classes, is pernicious in a country of this character, because there are not only different classes but different races. . . . To our mind, the only way that control could be effective would be to have all shops on estates, run directly by the estates.[14]

This planter suggests that European liberal doctrines ought not to be applied to the colonial context, where racial differences naturalize class differences and thus legitimate the strict surveillance and moral supervision of the working population.[15] As planters were afraid that eliminating liquor altogether would incite riots, the pragmatic solution was to have liquor shops run by plantations so that they could be controlled effectively.

In 1914, when the International Congress of Tropical Agriculture met at the Imperial Institute, South Kensington, under the presidency of Sir Ronald Ross, one of its major concerns was "tropical hygiene."[16] Ross suggested that the Congress should appoint a committee to consider the question of sanitation and hygiene on tropical estates. The *Planters' Chronicle* carried a report.

> Dr Harford (President of Livingstone College) said there were groups of men in all the different countries of Europe who were united to deal with the question of the supply of liquor to natives, which he believed had an important bearing on the efficiency of labour in the tropics. Some practical results had issued from the discussion of this subject at the last Congress. Belgium had enacted entire prohibition of the trade in spirits throughout the whole Belgian Congo. Portugal had taken similar action in the province of Angola, and France, Germany and Great Britain had taken strong steps in Africa during the last few years.
>
> Sir Sydney Olivier said that if agriculturists could be interested in the question of labour efficiency they would probably render great assistance to matters which had hitherto been regarded purely as

> problems of hygiene and medical aid. In all tropical countries he was convinced that a large amount of the apathetic disposition of labourers was due to latent disease which depressed their vitality.[17]

The rhetoric of drunkenness and disease was thus still very much available in the early twentieth century, when speaking of a native laborer whose difference from his employers was marked by both race and class. At this time, competition for labor was particularly strong, as many parts of south India were being developed as labor supply areas for other parts of the empire such as Ceylon, Malaysia, and the West Indies. Planters in India raised a strong protest with the government, arguing that they needed all the indigenous labor and could not afford to lose it. In one speech about emigrant labor, a planter suggested that these colonies were stealing Indian labor because Indians were essentially easier to discipline.

> The reason for the Indians being preferred to all others is undoubtedly that they are more docile and amenable to discipline, whereas John Chinaman or a Javanese—as a labourer—is a hard nut to crack and successfully deal with. As an artisan he is a most excellent workman and it is probable that if properly handled he might prove a valuable agriculturist. But his ways are peculiar and the western mind has not yet fathomed him nor worked in unison with him in the field. The result is that he is not so much sought after as the milder Indian, who can be brought under on easier terms, albeit the worm turns sometimes.[18]

The rendering of difference as inherent, residing in colonial natives' racial identities, marked them as inhabiting a region in which modernity had no space, in which the ideals of liberalism would appear incongruous. As another planter put it, "A coolie, that is to say, an Indian labourer, is a child in his ways; if you treat him kindly you can do almost anything with him."[19] This was, again, the rhetoric of feudal paternalism, not of the scientistic "labor power," defined as a measurable form of energy output that made all bodies equivalent. The body of a native laborer was not equivalent to that of either a white worker or a white *dorai,* for it was irreducibly determined by race.

It was not just an accidental lag in history that had the colonies following such an outdated discourse of labor through the first half of the twentieth century; the persistence of older (feudal, precapitalist, or premodern) forms of production and representation were necessary parts of colonial production. The combination of scientific discourses of progress with moralistic discourses of idleness and savagery might have seemed incoherent from the point of view of its contemporaneously evolving European social modernity, but this was a necessary, or functional, incoherence, as the moral discourse of work facilitated the operation of the scientific and economic systems of production. It is this functional incoherence that constitutes a "mixed" modernity. Postcolonial modernity incorporates both a mythos of techno-scientific progress and the specter of feudalism, not because the colonies could never quite get modernity right but because colonial modernity was necessarily constituted that way. I will return to this point at the end of this chapter.

Nature, Natives, and Modernity in the Plantation Economy

In the late nineteenth century there was a growing international demand for tropical products. Daniel Headrick, in *Tentacles of Progress*, his 1988 history of technology transfer and imperialism, characterizes the colonial production of such commodities as sugar, coffee, tea, cotton, quinine, and rubber as embodying the "new imperialism," which he contrasts with the "old imperialism" of the sixteenth and seventeenth centuries. Under the pressure of the needs of industrializing Europe, the new imperialism had to develop more efficient methods of producing and marketing tropical products than the old, which had merely "used the human resources of the tropics to collect the vegetable and mineral ones."[20] Headrick suggests that in the colonial era, two types of tropical agriculture, namely, peasant agriculture and plantations, emerged as the solutions to the problem of dramatically increasing the supply of crops for the world market. Both were subjected to intense study, experiment, and systematic modern management by colonial states—Dutch, French, and British. Headrick suggests, in a survey of these colonial powers' production of botanical commodities, that biology, economics, and politics were interlinked in the production of tropical export crops through both agriculture and planting.[21]

Headrick's work comprehensively reviews the institutional and economic background to colonial science. Although he argues that his focus is on the links among economics, politics, and science, he adopts a view that sees these as distinct spheres of operation, brought into interaction as a result of the political appropriation and manipulation of scientific and technological power. This separation of spheres is evident in his presentation of the central argument of the book, which is that although European technologies were successfully transferred to the tropics during the new imperial period (late nineteenth and early twentieth centuries), this geographical diffusion was not accompanied by a corresponding "cultural diffusion." The reason for this lag in cultural diffusion, he suggests, is that Europeans failed to school colonial subjects in the culture of science, while the colonized populations themselves were unable to understand how to use the new technologies properly. Thus the technologies of imperialism themselves are not examined as socially constructed artifacts but are seen as being used correctly or mistakenly according to the political and economic motives of their users. As Headrick describes it,

> The main argument of this book lies in the contrast between the successful relocation of European technologies under colonialism and the delays and failures in spreading the corresponding culture. The cause of this contrast was the unequal relationship between the tropical colonies and their European metropoles. In order to obtain the full benefit of Western technology through its cultural diffusion, Africans and Asians had first to free themselves from colonial rule and then—a more arduous task—learn to understand, and not just desire, the alien machinery.[22]

The notion that delays, and failures in the complete transfer of technology to the developing world account for the cultural and technological lags in postcolonial modernity underlies Headrick's history, as it does much scholarship on the history of technology in the non-Western world. As I have suggested, the mixed model of modernity gives us a much more accurate, fine-grained explanation of the "failure" of technological modernity, and better explains what is otherwise seen as contradictory postcolonial attitudes of incomprehension, envy, and desire toward technology.

Headrick assumes that the technologies as such were intrinsically beneficial. But the technologies in his book—ships, railways, telecommunication, sanitation, irrigation, economic botany, mining, and technical education—are as much political as technological artifacts, and cannot be studied as separable from their geopolitics. Headrick does not analyze the spheres of politics and economics as constitutive of technological artifacts; rather, they operate more as factors that inhibited or prevented the proper operation of technology.

Let us, then, examine both the discourses and practices, the texts and contexts, of the plantation economy. The scope of this section is, of course, not of the scale of Headrick's book, but I will nevertheless argue that the methodological lessons of this analysis will lead us to quite different conclusions about the specific relationship between colonial societies and imperial technologies.

In this chapter I look at the activities of planters in the southwestern hills of India during what Headrick characterizes as new imperialism, the late nineteenth and early twentieth centuries. I study planters' attitudes toward nature, natives, and scientific modernity, with a view to analyzing the complex and contradictory strands, or the mixed constituents, of colonial modernity in India.

Planters and Nature

The planters' relationship to nature was considerably different from that of the foresters. Where the foresters usually gave an æstheticized, almost metaphysical, priority to wild landscapes and the natural beauties of the untamed tropical jungle (even while taming this jungle into orderly "normal" forests), the planters were, as a body, generally more inclined to see tropical nature as primarily an untapped source of wealth.[23] Consider, for example, the following statement by the director of the Peradeniya Botanical Gardens in Ceylon, in 1909.

> The great development of European planting enterprise in the more civilised and opened-up countries has of course quite revolutionised the primitive agriculture or rather has built up a modern agriculture beside it. . . . [T]he northern powers will not permit that the rich and as yet comparatively undeveloped countries of the tropics should be entirely wasted by being devoted merely to the supply of the food and clothing wants of their own people, when they can also supply the wants of the colder zones in so many indispensable products.[24]

Although eighteenth-century botanical gardens had been areas of domesticated nature to which the leisured classes would come to admire the beauties of the plant world, by the late nineteenth century they were powerful vehicles of economic production, and directors such as this one had more at stake than the mere contemplation of the hundreds of acres of nature under his control. Nature (especially tropical nature), for the administrators of colonial botanical and agricultural research institutes, and for planting entrepreneurs at the end of the century, often represented an untold source of wealth. The gaining of this wealth was regarded as hard work, for one had to prevail against the climatic and topographical challenges involved in carving out extensive enclaves of tidy, regulated nature from a wild and chaotic tropical landscape. The professed motivation for this work, however, was not an emotionally charged environmentalism, as in the case of the early foresters, but an openly acknowledged desire for profit. In pursuit of this goal, many Englishmen moved with their wives and families to remote, hilly forested regions of the subcontinent to spend the major portion of their lives in virtual isolation from the rest of the country, managing (with nearly unlimited powers) hundreds of natives.

Although their primary motive was profit, nature was never, of course, absent from their experience. On the contrary, nature was everywhere: in the ground underfoot, rich in potential but thickly forested and in need of harnessing; in the vegetation around, which had to be cleared to make room for plantation residences and their English flower gardens; in the climate, with its fierce monsoons, leeches, and tropical insects. Nature was made amenable to the planters in many ways besides the ordering of the plantation sites themselves. A particularly intense aspect of planters' interactions with nature was in their recreational activities, which mirrored their professional activities in form, in that they carefully ordered nature to specific ends.

The institution that best embodies the planters' relationship with nature is the Planter's Club: a carefully tended oasis in the jungle to which planters from all over the area would come for sporting and social activity. Hunting, riding, and fishing brought planters into an interaction with nature that was commonly cast in terms of both romantic communion and masculinist domination. The club, in which membership was restricted to whites, sustained racial barriers and provided a site for upper-class English social relations to reproduce themselves through sports, games, and dances.[25] At the club, visiting government officials would enjoy the famed hospitality of planters, and the social relations that grew through these interactions contributed to the creation of a shared cultural context between planters and government servants. During the labor investigations at the turn of the century, a magistrate commented on the difficulties of putting these social ties aside when the interests of the planters conflicted with the regulations of the government: "peculiar difficulties beset a [magistrate] member of a small European community in a distant land who may be playing polo or billiards with a planter one month and the next month be asked to punish him for cruelty to a contract labourer."[26]

Despite their antagonism to the liberal currents in late nineteenth century British labor politics, planters gained an almost mythic status in English culture. A July 1899 issue of the *London Daily Mail* carried an article by an

English planter in India, in which he said, "The history of tea-planting in India and Ceylon is a splendid story of indomitable pluck. . . . The British planter in the East has literally made two blades of grass to grow where one grew before, he has reclaimed desolate mountains and inhospitable forests; he has provided work for hundreds of thousands of labourers, and he has poured new revenues into the public exchequer."[27] The article continued, describing the ardors of plantation life: "And the life the tea-planter leads—what is it? . . . in the outlying districts, where for weeks the planter never sees a white face, nor hears the sound of his mother-tongue, the loneliness is terrible."[28]

A BBC talk show in 1947 interviewed C. P. Goldsbury, a planter who had been assistant manager of an estate in Munnar in 1906. In the BBC account of Goldsbury's experiences, the planters come across as a band of pioneers who brought civilization to the outer reaches of the earth.

> He had seen the evolution of planters bathing out of kerosene tins to the snow-white enamelled slipper baths in their comfortable bungalows of today. . . . He was in Munnar before any roads were built, and the only means of communication between gardens were bridle paths. Every planter rode a horse and used to return from the club with a lantern tied on the stirrup, praying he would not meet a tiger, wild elephant or bison on the way. . . . One never shaved in those days, and if by chance a woman paid a visit to the garden, one generally dived into a tea bush. . . . Yes, tea is one of the richest heritages Britain will leave to India. It is an industry built by individual care to every single tea bush, cared for as children."[29]

The masculinist picture of the planter as venturing to places devoid of the domesticating touch of Englishwomen, and the representation of the planter's relationship to his tea bush as one of paternal care, reveal the plantation as a frontier site that nurtured the construction of masculine colonial identities. The paternalistic representation of the tea bush as child is, in fact, also a projection of one facet of planters' relationships with their native labor force. Planters commonly represented their laborers as children in need of a firm hand, whom the discipline of work would raise from their primitive state to true subjects of the empire. Although this was the dominant representation to the outside world, however, within planting communities the open antagonism between planters and their workers was commonly acknowledged. Instances of conflict were, in fact, often relished as occasions for planters to display their powers of controlling disobedient natives.

Planters and Native Labor

The most regular topic of conversation in club houses was the issue of native labor: how to procure it, how to keep it, and how to manage it most profitably. *The Planters' Chronicle*, a periodical which, in addition to reporting on issues related to planting, often carried planters' amateur verse for the amusement and entertainment of their community, carried the following verse on "the labour question," after noting that

> It is not difficult, at any time, to realise that this subject [labour] forms the most important topic of the planter's conversation, it crops up in his Planter's Association minutes, it confronts him on his daily round, it, in fact, dogs his footsteps by day and haunts his pillow by night.
>
> *The Labour Question Solved*
> —by Shakespeare Jr.
> If a cooly gives you cheek
> And you can catch the giver,
> To make him civil heave a brick,
> And dislocate his liver.
> When he refuses labour
> Or works in a manner airy,
> Just knock him down, and while he's there,
> Jump on his little Mary.
> And any time you find him
> Do what he didn't oughter,
> Dont hesitate, but shove him in,
> A pot of boiling water.[30]

Although this little piece should perhaps not be taken literally to indicate labor management techniques, it illustrates the violently antagonistic relationship that obtained between planters and their labor.

The verse expresses the frustration planters felt at the rate at which their workers escaped from plantation sites. Throughout the late nineteenth century, planters complained to the government that they did not have the laws they required in order to prosecute coolies (workers) who left the plantation while still indebted to the plantation owners and labor contractors. The India Act XIII (1859) and the Madras Act V (1866) had provisions to enforce labor contracts and regulate emigration of labor from Madras to neighboring princely states of Coorg, Mysore, Travancore, and Cochin, but the planters

FIGURE 7. Disciplining Labour. From Michael Edwardes, *British India 1772–1947: A Survey of the Nature and Effects of Alien Rule* (London: Sidgwick and Jackson, 1967), 329.

argued that they had not been adequately stringent. Their perceived need for more control over labor prompted the planters to form a representative body, the United Planter's Association of South India (UPASI), in 1893.

In March 1892, the District Planter's Associations of South India had submitted a memorial to the Government of India complaining about the labor regulations that obtained at the time under Act XIII of 1859, arguing that the act was designed for urban conditions and ought to be retailored to address plantation labor relations, on the basis that planters did not have access to labor as readily as urban employers, and hence ought to be given more power to keep laborers on their plantations. In 1896, the Government of India appointed a committee of enquiry into the subject.[31] The 1903 Madras Planters Act, which strengthened the system of penal contract, was an outcome of lobbying by UPASI and other Indian planting associations. P. H. Daniel characterizes the act thus: "Section 25 provided that if any worker absented himself without any sufficient reason from work or neglected his duties or refused employment he would be liable to pay a fine of 4 annas per day, for such period of absence or neglect and also to undergo imprisonment for up to 14 days. . . . Section 26 provided that when an estate manager complained in writing to a magistrate that a worker had run away, the magistrate should arrange for the production of the worker before him, by issue of summons or warrant of arrest. . . . Section 27 provided that any worker who ran away during the period of contract would be liable to pay a fine of Rs 50 or suffer imprisonment up to one month or both."[32]

If the so-called labor problem could be solved through force and the law, there remained other challenges which required recourse to new kinds of control. New levels of predictability and profit, the planters were discovering, could suddenly be mobilized, but only through the assiduous application of emerging forms of scientific knowledge.

Science, Global Trade, and the Planters

Science was a vexed issue for the early-nineteenth-century planters, who considered themselves more in the mold of fortune-seeking adventurers than of educated scientists and were therefore suspicious of science. However, the increasing importance of botanical gardens and agricultural experimentation led planting associations to pay systematic attention to scientific developments in seeds, climatology, hygiene, entomology, and soil analysis. UPASI would often appoint delegates to the meetings of the International Congress of Tropical Agriculture, and report activities of the Imperial Institute and other scientific institutions. The subject of "labour organisation and supply in tropical countries," in fact, often featured on the list of topics discussed at scientific meetings.[33] UPASI also put pressure on the government to begin agricultural experiment stations, and hire trained mycologists and entomologists.[34] It endowed a research position at the Calcutta School of Tropical Medicine, which was to pursue lines of research prescribed by UPASI. In 1920, the *Planters' Chronicle* carried an article on "The Farmer and the Expert" which suggested that "in planting, as in everything else, knowledge is power, and if the Planting Community have not that knowledge they must pay someone else to put this knowledge at their disposal. I trust I will not be mistaken, and be thought to mean that planters should themselves acquire a knowledge of science: . . . [but] if they have not this knowledge, they must obtain it from someone else."[35]

The following 1929 speech by Lord Melchett, identifying India as an important supplier of the empire's resources, sums up the modes of intersection of the science, commerce, and politics of planting.

> [T]he conception of the whole British Empire as a single economic unit is absolutely essential and urgent if the Empire is to continue at all. If we visualise the Empire and think of the magnitude of its territories, the wealth of its natural resources, we must be struck by the thought that the potentialities of the future must transcend both its glorious history . . . and its strong and vigorous present. If we consider its resources in minerals, in food products, in timber, in rubber, and practically every great commodity, in some of which, like nickel and jute, it enjoys a virtual monopoly, we find an economic complex so overpowering that there is no other economic unit which comes into the picture at all.[36]

The link between colonial natural resources and global trade was a central source of economic power for the British Empire. This aspect of planters' attitudes toward nature was a reflection of the larger business community's perceptions, in which nature presented infinite potential commodities. Melchett goes on to describe the necessary business arrangements.

> When we regard the whole problem from the right angle of a single economic unit, it instantly becomes obvious that to be a member of the British Empire carries with it absolutely concrete business advantages which nobody other than a member of that association is privileged to enjoy. Moreover, were the whole subject developed by business men of the Empire on modern business lines, those advantages are capable of a manifold increase. By arrangements of quotas and compensations taking due regard for what exists and visualising for the future, practical men can work out a practical scheme on merger lines. . . .
>
> India, in this development and expansion, in this picture of a gigantic merger of the whole British Empire, can play a role unsurpassed by any other part of the Imperial complex. Her vast territory, 1,805,332 sq miles. . . . Her natural resources of coal, iron, manganese, mica and other mineral and chemical raw materials are vast, as also of the tremendous range of her agricultural production. . . .

> Here, then is a basis for the implementing of a single Empire economic unit whose central idea would be the exchange of Britain's manufactured products for the agricultural products and raw materials of the various other component parts of the complex.[37]

Melchett hails the Linlithgow Commission Report, which recommended the establishment of an Imperial Council of Agricultural Research in India, as a spur to more scientific agriculture.

> Since the main exports of India are agricultural products like cotton. jute, tea, rubber, flour, rice, wheat, vegetable oils, hides and skins and wool, it is obvious that a thorough scientific policy of soil, seed and fertiliser research must enormously increase the value of her principal resources in the future. In addition, further big schemes of irrigation are in progress or under consideration, and much good work also remains to be done in putting Indian forestry on a systematic basis.
>
> India, then, as a supplier of Empire resources, is big with potentialities for the future. Imperial Chemical Industries, Ltd., for example, including Burma and Ceylon, is in close touch with the official Agricultural Departments in connection with experiments on all the principal crops and in particular on rice, cotton, sugar-cane and tobacco. . . . Scientific propaganda and research will be able to accomplish [much] in placing India in the foremost rank as a supplier of Empire resources.
>
> The great keys of modern scientific research and modern business organization are in India's own hands. . . . Her resources are great, and as a component part of the great Imperial complex her position is assured. She can play a leading part in establishing that real harmony between industrial and agricultural production which is essential to the prosperity and stability of the Empire.[38]

By the end of the 1920s, planting, like forestry, had entered a stage in which the union of scientific and business methods was aggressively pursued. The institutionalization of scientific research and management in both cases led to a pace of growth quite different from that of the earlier period. Nevertheless, the legacy of the earlier period lived on in the representations of work, racial difference, and a scientific modernity in which the term "India" referred to the unit of Great Britain that supplied her natural wealth. This continued to be a modernity whose contradictions abounded for its native subjects.

The Mutual Constitution of Science, Politics, and Culture: Toward a "Mixed" Model of Modernity

As I have suggested, the contradictions between scientific production for a global market and premodern labor patterns were not a result of a lag or oversight on the part of colonial officials, nor were they due to the inability of the natives to appropriate the tools of progress. The model of progress that saw India as the supplier of resources to the British Empire was the same model that required a constant supply of cheap, disciplined labor. The rhetoric of progress, and the accompanying hierarchy of savage to civilized societies that was repeatedly invoked to justify coercive labor practices, were not epiphenomena but functional aspects of this system.

Linear models of scientific modernity that postulate the supersession of older, premodern modes of discourse and practice by newer, scientific modes neglect the ways in which the old modes are in many ways constitutive of the new. Periodizations such as Rabinbach's capture the complexity of neither the processes of change nor the resulting set of practices we call modernity. Models such as Headrick's are inadequate in explaining the ways in which native identities and the promise of technology were themselves constructed and represented in ways particular to the moral and productive discourses and practices of the time.

In order to illustrate how commonly and unproblematically a separation of spheres (such as politics, economics, science, and culture) is assumed, consider the following article on "Science and the Nineteenth-Century Ceylon Coffee Planters." Historian T. J. Barron, in this article, assumes that the spheres of morality, science, politics, and economics operated separately and distinctly in the nineteenth century. He argues, "For British advocates of empire, the claim of the imperial power to superior scientific understanding was seldom of first importance. It was *morality, not science,* that was held to justify empires."[39] He goes on to acknowledge that the creation of the Board of Agriculture (established in 1794) and the Agricultural and Horticultural Society of Calcutta (established in 1824) were important landmarks in scientific and technical assistance to planters; that botanical gardens played a large part in transplantations and the accompanying expertise; and that after 1850 there was considerable investigation into microclimates, soils, and hybrid seeds. Nevertheless, he asserts, religion spearheaded the endeavor, while science played an important but separate part. "If God was unequivocally the manager of the colonial enterprise, it was science which captained the team." He erects a similar dichotomy between science and economics: "For the Europeans to set up their estates was a major development, involving the provision of large amounts of capital, the supply of seeds or plants, the creation of buildings, the gathering of a labour force, and much else. But these were chiefly *economic* issues. *No new scientific knowledge* was involved."[40]

And there was a similar separation between the spheres of science and politics. "The ascendancy of the European planter over his local rival becomes, despite his protestations to the contrary, a matter *not of scientific knowledge* but rather of the obtaining of *political* favours, discriminatory policies on access to lands, markets, taxation, fiscal matters."[41]

Barron's article exemplifies the ease with which distinctions between economics, politics, morality, and science are made in historical analyses of a period that is crucial to the understanding of the roots of modernity. Models of linear causation that assume separable systems of politics, society, and science oversimplify the process by which these categories themselves are employed in interconnected constructions of each other, and are all themselves historically changing. By analytically separating these spheres, we gain local insights into some historical moments in the development of each, but no useful comprehensive model of modernity, in its complex totality, can be drawn in this way.

By identifying how interlocking elements (economics, science, religion) supported each other and an overarching vision of modernity, we can see the contest between precolonial tribal communities and colonial groups not as a showdown between the dark forces of irrationality and the enlightened rationality of scientific knowledge but as an historically situated process of conflict between two rational systems of knowledge about nature, undergirded by unequal structures of economic, political, and social practice. In the next two chapters I continue to argue for the complication of our models of modernity, adding two more elements to the stew of "mixed" technoscientific modernity.

Chapter 5

Ethnographers

[I]f the strangers who are wandering aimlessly about and dying on the outskirts of the town should turn out to be Yenadies, that will mitigate in the same measure the nature of the calamity. For the lives of wild and unreasonable men of the jungle are not to be regarded as of equal value with those of industrious and satisfactory villages, nor can they be equally protected.[1]

In a political as well as in an epistemological sense, imperial ethnography contributed to the construction of natives—especially tribals—as natural resources. The last three chapters have indicated the importance, for the British, of knowing the natives: Ooty was settled by cajoling the Todas and Badagas out of their lands; the Madras forests were cleared and tended by the Kurumbas, Chenchus, and numerous other tribes; and the plantations of south India produced a steady flow of coffee and tea for export as a result of a large, practically enslaved, labor force of landless peasants and tribals. This chapter explores the paradigm that made it possible for the colonizers to know the natives: the scientific paradigm of anthropology, the "science of man."

This chapter has five sections. In the first, I briefly review the disciplinary basis of colonial anthropology, and its status within the academic field of British anthropology in the late nineteenth century. In the second section of this chapter, I explain the significance, in the colonial context, of ethnographic ways of knowing, through examining the discourse and practice of anthropology in south India in the late nineteenth and early twentieth centuries. In the third section, I assess the political significance of the epistemological framework that ethnographic knowledge made available to colonial administrators and scholars. The fourth and fifth sections of this chapter are exploratory. They begin to investigate the dynamics of indigenous responses to colonial anthropology, while acknowledging that this is a subject complex enough to merit a separate study.

Colonial Anthropology

Over the past two decades, academic anthropology has, as if awakening from a long slumber, discovered its roots in objectivism and imperialism, and turned to both history and theory in an attempt to come to terms with its origins. The skeletons in anthropology's colonial closets have been exhumed and minutely examined, and the analytical results hung, albatross-like, around the necks of twentieth-century anthropologists in expiatory rituals intended to cleanse the discipline for its new poststructuralist, postcolonial, postmodern, no-longer-Enlightenment missions.[2]

What, then, is the excuse for yet another academic analysis of the colonial history of anthropological knowledge? It is almost too easy to satirize British anthropologists in the colonies, invested as they were with an earnestness of purpose and a pioneering sense of forging a discipline as they crossed the boundaries of civilized men's experience. I would like to disclaim, at the outset, the intention to indict, retrospectively, the discipline for tainting its science with ideology.[3] Nor am I primarily concerned with separating the "real" anthropologists from the amateur ethnographic data collectors who had administrative day jobs; nor with chronicling the writings of the major colonial anthropologists of the late nineteenth century. This chapter attempts to put the practice of anthropology in the context of other sciences of natural resource use, and to explore the concurrent social meanings of ethnographic ways of knowing.

It might seem odd that an investigation into early anthropology should be justified on the basis of an interest in the history of natural resource use, when we think of anthropology today as a human or social science, clearly distinct from a natural science. The initial motivation to investigate anthropological knowledge arose for me, when, on investigating instances of scientific control and management of the environment, I found that foresters, planters, and other improvers of the colonial tropics repeatedly encountered the "problem" of the native user; that is, forests turned out to be composed not merely of vegetation and animal life but also of human life, and these human inhabitants appeared to use the forests actively for their daily needs. The ideal of scientific management of nature that late-nineteenth-century foresters cherished was rendered somewhat messy by the presence of these natives. An efficient system of management of tribals was called for. A survey of foresters' and planters' journals shows that both groups were avid collectors and readers of ethnographic information. Forest Department offices and forestry classrooms invariably made references to anthropological texts that dealt with the specific regions and tribes under consideration, and forest officers often submitted amateur but detailed pieces to journals that were clearly modeled on ethnographic styles of research. Toward the turn of the century, anthropology was increasingly turned to by administrators and by nongovernmental agencies, such as industrial undertakings and Christian missions, as a systematic way of understanding nonindustrial societies.[4]

Anthropology's close connection with the nineteenth-century growth in systems of natural resource use is confirmed if one looks at the early stages of the history of anthropology as a discipline.[5] Before it received formal recognition by the British Association for the Advancement of Science, anthropology had in fact been classified along with natural history.[6]

Historian Henrika Kuklick explains that even though it was only formally recognized as a science in 1884, anthropology, perceived as the pursuit of useful knowledge about the resources of the non-Western world, grew through the nineteenth century. Its intellectual roots drew upon classics, biblical studies, and philosophy, but it was best appreciated as a type of natural history, one species of a class of knowledge—which also includes geology, botany, zoology, and geography—that could be used to manage the nation's resources of people and land.[7] Thus British anthropology in the nineteenth century was explicitly—theoretically and practically—connected with the study of the natural resources of the colonized world.

Kuklick sees the development of anthropology in the late nineteenth century as propelled by a Utilitarian belief in the boundless potential of science. Geology, exploration, and anthropology were linked both institutionally and intellectually. Naturalist T. H. Huxley, who worked for the Geological Survey, was president of the Biology Section of the British Association and of the Ethnological Society of London, and campaigned vigorously for the establishment of the Anthropological Institute. He argued that such social research was "the application of methods of investigation adopted in physical researches to the investigation of the phenomena of society."[8]

Kuklick describes the trends in anthropology as an academic discipline—evolutionism, diffusionism, and functionalism—that came to dominate academic anthropology at different times during the late nineteenth and early twentieth centuries. Significantly, they all appealed for government support and recognition of anthropological work on the grounds of its utility to colonial administrations. In addition to Colonial Office officials, leaders of missionary societies and wealthy philanthropists became patrons of anthropology for its capacity to "do good and to create a stable world order in which international trade could flourish."[9] Kuklick suggests that the discipline's constitutive phase occurred in a period when Britain was becoming an increasingly powerful global force because of its imperial territories, and that the "institutional insecurity of the discipline" in this period mirrors the "precarious authority of British rule in much of the Empire." In their struggle for professional legitimacy, anthropologists promoted themselves with arguments that dovetailed with the liberal apologist argument for imperialism, which went roughly as follows: "self-interest required Britain to act to preserve traditional cultures, for in those tropical areas that were unsuitable for European settlement, areas that constituted a considerable portion of the Empire and housed those backward societies particularly vulnerable to disintegration, peoples could not serve as labourers in the economic enterprises of Empire unless their cultures remained viable."[10]

The argument that worked best, Kuklick reports, both in the discipline's campaign for professional recognition in England and in its attempt to gain legitimacy as a practical science in the colonies, was "that anthropology was value-neutral, technical expertise, grounded in scientific method, enabling practitioners to identify causes of social conflict and formulate solutions equitable to all parties, thus transforming political issues into administrative ones."[11] This argument is especially relevant to our understanding of anthropology as a solution to the problem of unruly native inhabitants of forests. If tribal use of forest resources had been seen as a political issue, all manner of complex discussions regarding rights, responsibilities, and democratic representation would have had to be developed around this issue. By naturalizing rather than politicizing tribal practices, a scientized system of knowledge was developed whereby conflicts over different modes of utilization of forests were recast in terms of scientific (inherently progressive) systems of knowledge versus unscientific (inherently backward) systems of resource use. Once this had been established, the problem was then literally redefined as an administrative problem; colonialism's burden of bringing modernity to the primitive world could be seen as merely a question of systematically stamping out persistent nonscientific practices through the imposition of an ordered set of practices that together defined a rationalized, progressive use of nature. Thus the establishment of a scientific regime of natural resource management was dependent on a specific kind of ethnographic knowledge, namely, one that could translate its knowledge of natives into systems of documenting and controlling tribal populations' ways of knowing and using their environment.

Kuklick explains why such a scheme seemed ideally suited to the colonial context. "The objectives of colonial rule seemed ideally realized through the application of scientific principles. Indeed, . . . many British social reformers considered colonial regimes exemplary, both because they had replaced politics with administration and because they selected and rewarded their staffs according to meritocratic standards—the values dear to the new middle class from which so many anthropologists and colonial officials were drawn."[12]

Kuklick's interest is in the historical development of the academic field of anthropology, in which she sees colonial anthropology as a powerful but ultimately marginally influential episode. Thus, she concludes that since colonial governments repeatedly contrasted "practical" with "academic" knowledge, and often did not change their policies as a direct result of professional anthropological studies, we can see that "No matter how well funded they were, . . . academic anthropologists remained suspicious characters to colonial regimes. . . . [C]olonial civil servants were unwilling to relinquish the management of the Empire to scientific experts."[13]

As an academic discipline, professional anthropology—despite its occasional successes with colonial technocrats and grant-giving bodies—was ultimately unable to ensconce itself comfortably in the offices of imperial power. Kuklick's judgment about the significance of colonial anthropology is correct insofar as she assesses its influence in terms of its effects on mainstream British anthropology understood as a profession, defined by journals and academic reputations back in its metropolitan center, and in terms of professional anthropologists' success in achieving funding and official posts in colonial administrations. I would, however, like to take up an avenue that she refers to only briefly: the importance of practical anthropology, practiced by colonial officials who were considered amateurs by academic anthropologists. These amateur anthropologists (revenue or census officials, museum superintendents, foresters, missionaries, or writers of gazetteers in their official capacities) collected and published what we might call ethnological or anthropological information in various contexts. Charles Morrison's 1984 article, "Three Styles of Imperial Ethnography," surveys this sort of research, and categorizes it into official revenue literature, quasi-official compendia and miscellanies, and officially subsidized tribal monographs. Although none of these categories falls under what Kuklick would call academic British anthropology, Morrison argues that this ethnographic research, produced by colonial officials, "is still of some ethnological interest, and . . . had a definite . . . connection with developments in academic anthropology."[14]

My interest in this research, however, is slightly different from both Kuklick's and Morrison's, in that my aim is neither to give a comprehensive history of the academic discipline of anthropology nor to assess the ethnological value of specific treatises. Rather than respect the lines between academic anthropology and messy colonial politics, I would like to study the knowledge that was produced at the intersection of the academic discipline and colonial society. That is, I am interested in how ethnographic ways of knowing made their way into the thinking and writing not just of professional academics but also of the sort of colonial person with whom a tribal or peasant community might have had more frequent and sustained interactions: missionaries who sought to uplift them, foresters who sought to protect the forests from them, planters who sought to settle them in cooly lines, and rural administrators engaging in amateur ethnography.

What does the accumulation of knowledge about subject peoples in the colonial context tell us about the ways in which sciences and colonial society constituted each other? Specifically, how was the colonial native constructed as an object of scientific knowledge? If anthropology was a scientific enterprise continuous with geography and natural history, were these processes analogous to the ways in which other colonial resources were drawn into a scientific system of management and utilization?

Ethnographic Ways of Knowing

Historians of anthropology have laid out in detail the nineteenth-century debates between polygenism and monogenism, Darwinism and Lamarckianism, and the transitions between evolutionism, diffusionism, and functionalism that took the discipline into the twentieth century. For the purpose of this chapter, what George Stocking has referred to as the late-nineteenth-century "classic" concept of race is most useful in describing the elements of the working model with which many amateur colonial ethnographers operated. At the turn of the century, Stocking argues, although distinguishable strands of the "antecedent traditions" of ethnology, Lamarckianism, polygenism, and evolutionism existed, several social scientists, and most lay people, effectively operated with some (not necessarily coherent) mixture of these intellectual traditions. Roughly, he suggests, a formulation of the "classic" concept might have looked like this 1895 statement by G. Stanley Hall, a leading American psychologist: "The color of the skin and the crookedness of the hair are only the outward signs of many far deeper differences, including cranial and thoracic capacity, proportions of body, nervous system, glands and secretions, vita sexualis, food, temperament, disposition, character, longevity, instincts, customs, emotional traits and diseases."[15]

In this context, race and civilization were commonly spoken of as linked in an evolutionary pyramid in which various human groups developed along trajectories that moved from the lower races (color-coded black, red, brown, or yellow) toward the social and cultural forms of the Caucasian races of northwestern Europe. Stocking discusses examples of the persistence of eighteenth-century ideas of civilizational progress (in terms of cultural evolution from savagery to civilization) along with a Darwinian racial evolutionism, a polygenist romanticism, a biblical ethnological framework, and a Lamarckian conflation of the social and the biological.

This pregenetic, late-nineteenth-century anthropological concept of race is commonly found operating either implicitly or explicitly in amateur, or practical, colonial ethnography. This chapter seeks to understand the effect of these practical ethnographic discourses on the day-to-day lives of tribals, in terms of the restrictions placed on them and the newly constrained survival options open to them. My examples are drawn mainly from the Nilgiri and Malabar Hill districts, but will be supplemented with examples drawn from the other hill regions of Madras in instances where a phenomenon was presidencywide, such as, in this chapter, the operations of the Criminal Tribes Act and the use of anthropometry to facilitate the administration of law and order.

South Indian Ethnography and the Significance of Knowing the Natives

The Malabar and Nilgiri Districts of south India were particularly attractive to ethnographers (both professional and amateur) for what they referred to as the diversity of native life in these districts. For example, Fred Fawcett, author of an ethnographic monograph published in the *Madras Government Museum Bulletin*, explained why anthropologists could collect valuable data in Malabar.

> Malabar is peculiarly interesting in the diversity of its peoples, for there are to be found Vedic Brahmins of the purest Aryan type in the South—the Nambûtiris—and people of every grade down to the wildest denizens of the jungles to be found in Southern India—the Ernâdens. Study of its forest peoples alone might well occupy more than the lifetime of one investigator. In it is the Jew—for we may include the Jew as an inhabitant, since he has perhaps lived in it as long as the Anglo-Saxon has in England, and adopted the vernacular—with a nasal index of 61.5, and the Ten Kurumbar of the Wynad

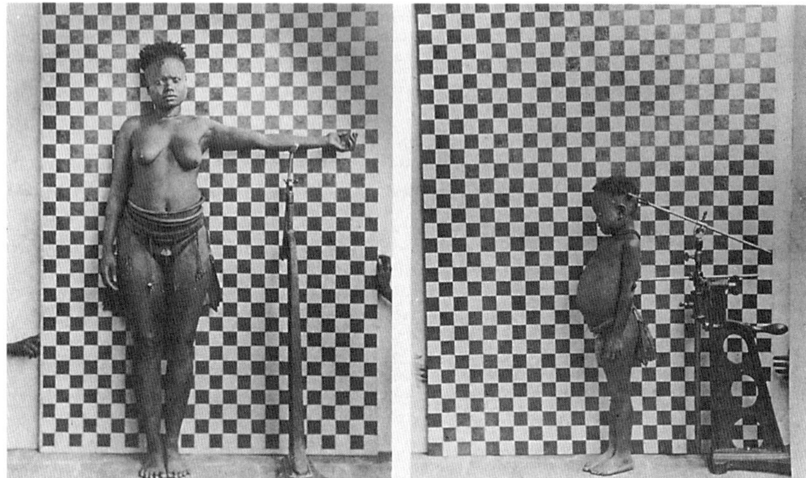

FIGURE 8. Photographs of Andaman Islanders taken by French photographer Oscar Mallitte in 1857. From Ray Desmond, *Victorian India in Focus: A Selection of Early Photographs from the Collection in the India Office Library and Records* (London: Her Majesty's Stationery Office, 1982), 55. *By permission of the British Library.*

> forests, whose nasal index in the case of the men is 96.8, and in the case of the women 100.7. And there are peoples whose index forms a graduated scale between these points of minimum and maximum. . . . Among the inhabitants of Malabar . . . are the most dolichocephalic people on record, the Mullu Kurumbars with narrow hands and feet as well as heads, and the Polayans, black in colour, and shortest of all the races of the world.[16] . . . There too, we meet with wide diversity in custom, polyandry, polygamy and monogamy; the system of inheritance through women.[17]

This quote is representative of late-nineteenth-century ethnography in at least two ways: first, its heavy reliance on anthropometric measurements for the classification of various tribes into a hierarchical pyramid, and second, its incorporation of Brahminical hierarchies into an anthropological, racialized civilizational framework of progress. Neither of these factors was confined to the colonial context; anthropometric data had been used to classify European into different racial strains, and the racial linking of Brahmins with Aryans was common in work on Indo-European peoples. Colonial ethnographers, moreover, strove to connect their data with academic theories of the time, and saw themselves as contributing to a pool of universalizable scientific knowledge—hence reference to established anthropometric texts was common.

The production of anthropometric data and its incorporation into accepted scientific theories helped to establish as true, among a wide—not exclusively academic—community, a hierarchy of races within the subcontinent, going from primitive dark-skinned jungle dwellers to the civilized (or civilizable), light-skinned Brahminical types. The identification of the Brahminical with the Aryan or civilized type gained, as one might imagine, considerable significance when articulated in a context in which scientific education was more easily available to Brahmins than to any other group.[18] Several south Indian Brahmins took up the scientific study of anthropology; the most notable among them, L. K. Ananthakrishna Iyer, produced, on the standard colonial model, several volumes of "Tribes and Castes" studies and numerous monographs on south Indian tribal societies.[19] By identifying themselves with the superior strains of human types, Brahmins were able to view other Indian castes and groups as falling within a paradigm that fit with a hierarchical reading of their own scriptures, as, for example, the model afforded by the *Manusmriti*. Further, by identifying their observations with the objectivity of modern science, they could believe themselves to be transcending the objectifying gulf between colonizer and colonized. This gulf was overcome not by deconstructing it but by transporting themselves to the other side by means of the "scientific method."

Academic anthropology in India today often unself-consciously sees itself as preserving this scientific tradition. A graduate anthropology text, *The Tribal Culture of India* (1977), explicitly draws its professional genealogy as follows.

> Man and society have been the subjects of study in India from time immemorial. The *Manusmriti* gives an exhaustive social and structural account of the people of India. The main principles of the social structure presented by Manu are many but mention may be made of some, like (i) the theory of origin of the universe, (ii) the principle of *Varna*, (iii) the principle of *Ashraṁ*, (iv) all human beings are born unequal in their capacity of execution but are all equal in their capacity of enjoyment, (v) that society must be based on division of labour . . . (ix) that no person should acquire wealth except for landed property by inheritance, and so on. In fact, he has given an eternal or *sanatana* core of the infra-social structure. . . .
>
> Valmiki . . . has also given an ethnogaphic account of Bharatvarsha

> by describing the Arya, Deva, Danava, Rakshasa, Vanara, etc. . . . Tulsidas and Kabir made a penetrating study of man and came up with substantive material . . . about the folk people of India in particular.
>
> In the modern period the study of man and his society became the subjcct-matter of anthropology. The study of man in its totality got a scientific footing through it. . . . [T]he establishment of the Asiatic Society of Bengal [1774] . . . marks the beginning of a scientific tradition in India for the study of "nature and man."[20]

Several thousand years of history and literature are teleologically telescoped into a history of science, with literary and religious texts shorn of their cultural and ideological context and made to stand in for proto-anthropological thought. Caste, the division of labor, and the restriction of property ownership to a privileged elite are naturalized as elements of an eternal social order, in which social groups are categorized with terms such as Arya and Rakhshasa, which had strongly racialized connotations in the nineteenth century.[21] Anxious to represent anthropology as a science in the late twentieth century, this textbook identifies anthropology's formal entry into the modern era with the establishment of colonial institutions such as the Asiatic Society. Outlining some of the major nineteenth- and twentieth-century publications, the introductory section concludes:

> Thus, Indian Anthropology, which was born and brought up under the predominant influence of British Anthropology, matured during the constructive phase on the lines of English Anthropology. . . . [T]he tradition of tribal studies as the exclusive focus by the enlightened British scholars, administrators, missionaries and the then British and Indian anthropologists continued till the end of the [1940s]. On the lines of Anthropology as taught at the time at Cambridge, Oxford and London, Indian Anthropology was characterized by ethnological and monographic studies with a special emphasis on researches in kinship and social organization.[22]

The preoccupation with the study of tribal customs was shared by colonial and indigenous ethnographers, fitting in as it did with both assumptions about universal progress from primitive to Western European types, and Brahminical models of the origin and organization of society.[23]

The same textbook specifically notes, as do most surveys of Indian anthropology, that the Nilgiris were the site of some of the most extensive and sustained ethnographic fieldwork in south India. The earliest British accounts of the Nilgiri tribes are two studies of the Todas by army officers stationed in the Nilgiris: Captain Henry Harkness's *Description of Singular Aboriginal Race Inhabiting the Summit of the Neilgherry Hills or Blue Mountains of Coimbatore in the Southern Peninsula of India* (1832), and Lieutenant-Colonel William E. Marshall's technical phrenological study of the Todas, *A Phrenologist amongst the Todas or the Study of a Primitive Tribe in South India, Their History, Character, Custom, Religion, Infanticide, Polyandry, Language* (1873). The Todas were the most studied group in the area, and were the subject of various theories identifying them as an ancient lost Jewish tribe, Sumerian survivals, Scythians, Druids, Romans, Asian Caucasic, Indo-Afghan or Assyroid mixtures, or pre-Dravidians.

Harkness's book is written in the form of a travelogue, whereas Marshall offers detailed tables and figures in support of his thesis that the Todas are an example of the most primitive form of humankind. He sees evidence of this in their "torpid and inefficient nature," an "aversion to all forms of *profitable* labour, and incapacity for commercial pursuit," observations for which he offers phrenological measurements as support.[24]

> The Toda is merely a simple, thriftless, and idle man, who will never, so long as his blood remains unmixed with that of superior tribes, or, by selection, is improved almost beyond recognition, work one iota more than circumstances compel him to do: but without taint of the ferocity of savagery.
>
> I now proceed to compare the known qualities of the Toda with the form of his cranium. . . . The Toda tribe is entirely, and without individual exception, narrow-long-headed—dolichocephalic—every person in it, of both sexes, being deficient in every organ at the sides of the skull; and . . . having the perceptive organs over the eyebrow . . . , and the Domestic group at the back of the head, large. If we add to these indications the deficiency in morals and in superior mental organization which appears to be an universal attribute of almost entirely undeveloped peoples, we can, I think, make up our minds without hesitation, as to what form of skull is the most primitive of which we have yet discovered remains.[25]

Marshall's writing is a mixture of polygenism, Lamarckianism, and evolutionism, and displays a close familiarity with the technicalities of phrenological measurement and calculation. The observation that is most central to his argumentation, however, is that the Todas practice no agriculture, and hence are lazy, thriftless, and backward. The ability to labor is represented as an inherent trait, and one that marks most clearly the distinction between primitive ("noncommercial") and advancing ("industrious") races. Marshall found the other tribal groups in this area more inclined to agricultural labor, more industrious, and therefore higher types than the Todas.

The most famous ethnographic treatise on the Todas is W.H.R. Rivers's *The Todas* (1906). Rivers was more inclined to see the Todas as degenerate elements of a higher race than as survivals of an ancient primitive race, as Marshall did. In their religion and their "aquiline" noses, Rivers found evidence of a "higher culture"; however, he, too, regarded their lack of building skills and disdain for manual labor as evidence of backwardness. Rivers hypothesizes the Todas to be "representatives of one or more of the castes of Malabar whose institutions have in some ways degenerated during a long period of isolation," or, alternatively, "one of the hill tribes of the Western Ghats who have developed a higher culture than the rest in the very favourable environment provided by the Nilgiri plateau." He thus saw the Todas as somewhere between their more primitive neighbors and the civilized higher-caste Hindus of the Malabar plains, and attributed their positive features to the favorable climate of the Nilgiris (which, as we have seen in chapter 2, was commonly likened to England's climate). He speculates that further research on "the existence of this strange people may help to illuminate the many dark places which exist in our knowledge of the connexion between the Aryan and Dravidian cultures."[26]

The best-known ethnographer of south India was Edgar Thurston, superintendent of the Madras Government Museum. In his official capacity he traveled extensively throughout Madras Presidency collecting ethnographic observations, which he reported in regular issues of the *Madras Government Museum Bulletin*. His seven-volume *Tribes and Castes of Southern India* (commissioned by the Government of India's Ethnological Survey in 1901, when it appointed superintendents of ethnography for each province), alphabetically surveys every tribe and caste group in the south. Much of his field work was done by his "assistant" K. Rangachari, who acted as interpreter, photographer, and recorder (on phonograph) in Thurston's presence, and conducted field work on his own while Thurston was away in England.[27]

Thurston also wrote the volume on Madras for a series on provincial geographies of India, in which he combines a textbook-style account of physical and cultural geography with ethnographic detail on habits, customs, and festivals of the natives. Ethnographers were the crucial link, the knowledge translators, between the rulers and the ruled. The editor's preface to the vol-

TABLE 1. *A Tabulation of Nasal Indices for South Indian Tribes and Castes*

Tribe name	Linguistic area	Nasal index	Structure
Kurumba	Canarese	74.9	162.7
Boya	Telugu	74.4	163.9
Tota Balija	Telugu	74.4	163.9
Agasa	Canarese	74.3	162.4
Agamudaiyan	Tamil	74.2	165.8
Golla	Telugu	74.1	163.8
Vellala	Tamil	73.1	162.4
Vakkaliga	Canarese	73	167.2
Dasa Banajiga	Canarese	72.8	165.3
Kapu	Telugu	72.8	164.5
Nayar	Malayalam	71.1	165.2

Source: Extracted from a larger table in Edgar Thurston assisted by K. Rangachari, *Castes and Tribes of Southern India,* vol. 1 (1909; reprint New Delhi: Asian Educational Services, 1987), li–lii.
Note: The table is followed by the text: "This table demonstrates very clearly an unbroken series ranging from the jungle men, short of stature and platyrhine, to the leptorhine Nayars and other classes."

ume makes an argument for the political value of ethnographic knowledge: "the success of British rule in India is largely due to the fact that the early administrators adopted the local systems of government and moulded them gradually according to the lessons of experience. And this was because the British occupation was that of a trading company of which the present Government of India is a lineal descendant.[28] Here he identifies successful administration with accurate and reliable knowledge about local custom, and credits the original recognition of this to the commercial motivations of early British occupiers of India. Further, with the recent institutionalization of provincial legislative and executive positions, he sees an increased need for detailed, specific knowledge of each region of the country and suggests that this is where ethnographic techniques are most valuable.

> Endeavours have been made to select as authors those who, besides having an accurate and detailed knowledge of each area treated, are able to give a broad view of its features with a personal touch that is beyond the power of the mere compiler. . . .
>
> Mr Edgar Thurston . . . [was] Superintendent of the Madras Museum for twenty five years. . . . [H]e sampled every form of natural product in the south. As Superintendent for many years of the ethnographic Survey he travelled through every district and obtained

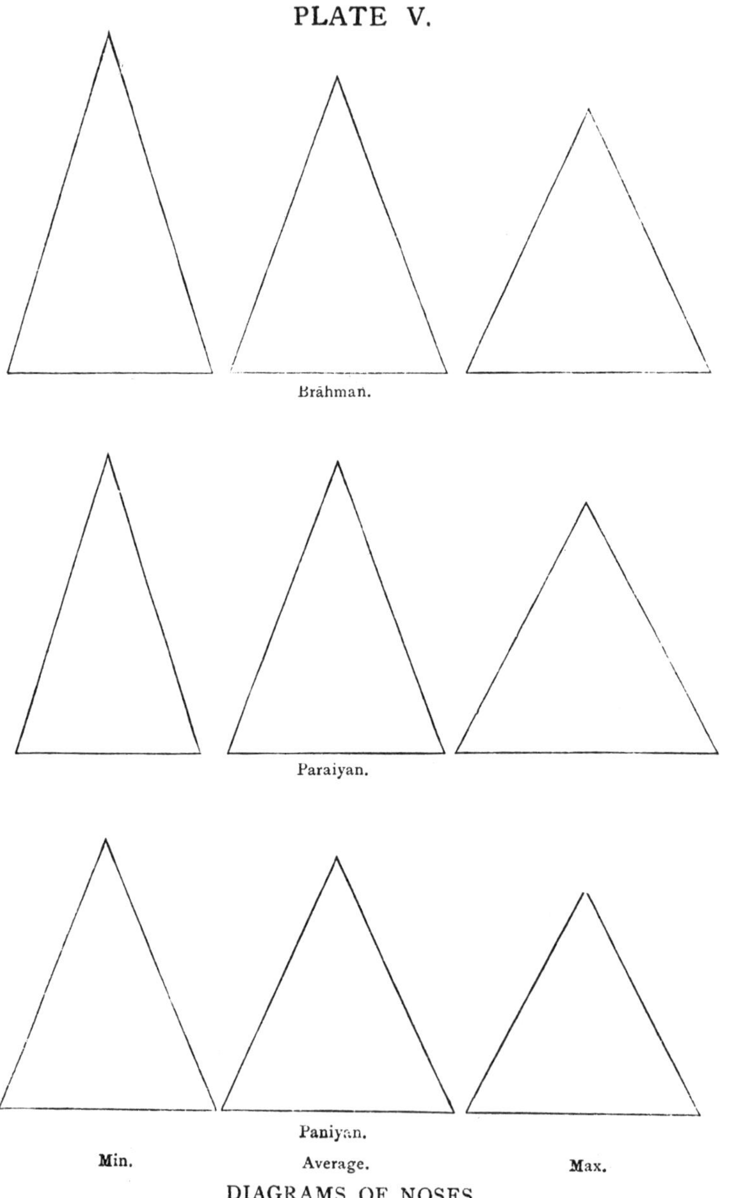

FIGURE 9. Diagrams of Noses. From Edgar Thurston assisted by K. Rangachari, *Castes and Tribes of Southern India,* vol. 1 (1909; reprint New Delhi: Asian Educational Services, 1987), iii.

> an intimate acquaintance with the people, his numerous publications on ethnography being summed up in his encyclopaedic work on the Castes and Tribes of Southern India. . . . [N]o one else could so readily recall an appropriate story or legend to add to the human interest of nearly every place mentioned.[29]

Here we see the link between administration and ethnographic knowledge made explicit. Thurston's "personal touch" yields a type of knowledge whose power is greater than the sum of its compiled facts, the editor tells us. This personal knowledge has qualified him to write an account commissioned to facilitate the administration of the country. The characteristics of ethnographic knowledge—its personal acquaintance with people and customs, its vast store of local experience, and its methods of combining systematization of these data with personal, anecdotal evidence—provided an important framework by which to conceptualize a form of government that sought to represent itself as a centralized power with localized expertise, firm yet understanding, authoritative yet knowledgeable. Thurston's wide experience with the tribes of India is characterized as an exhaustive knowledge of the "natural products" of the south.

Thurston's work was part of a nationwide effort to collect technical knowledge of all the major tribes and castes of the Indian empire, a project which, more than any other kind of practical colonial ethnography, bridged the gap between academic and colonial anthropology. Some of the officials involved in this project submitted their resulting publications for degrees in anthropology in England, and several became professors in Cambridge and London.[30] The Ethnological Survey of India was probably the most ambitious project of colonial anthropology, and was drawn up by Herbert Risley, census commissioner in 1901. Risley's blueprint for the survey included elaborate notes in ethnographic methodology, and a standardized questionnaire on the basis of which provincial officials were to collect information. The project was never actually completed, though Thurston's response was the most comprehensive.[31]

In this project, it is possible to discern the connection between administrative policy and anthropological ways of knowing on the one hand, and the construction of the native as a natural entity, on the other. Technical knowledge about the native could qualify one to move between the spheres of government and academe. And, in fact, tribals were constructed as objects of knowledge not only by anthropologists but also by missionaries, educationists, and the police.

Missionaries, Morality, and Ethnographic Ways of Knowing

Lieutenant-Colonel W. E. Marshall, the amateur ethnographer we encountered in the last section, had been introduced to the Todas by Reverend Friedrich Metz of the Basel Mission Society. Marshall reports that Metz was the only European to speak the Toda language, having "spent upwards of twenty years in labours amongst the primitive tribes" of the Nilgiris. On furlough in Ooty, Marshall had "long been curious to understand the mysterious process by which, as appears inevitable, savage tribes melt away when forced into prolonged contact with a superior civilization," and had set about studying the Todas in order to answer this. He hoped, through this study, to demonstrate the utility of phrenological techniques to ethnography. He believed he could see "much of the blameless Ethiopian," and "something of the Jew and of the Chaldean" in Toda appearances, and set out to uncover physiognomic proof of their primitive natures.[32]

Marshall's initial hypothesis links him with the biblical ethnological tradition that was a strong academic school of thought in Britain until 1860. Marshall's, and many others', writings on the Todas closely followed the framework of biblical accounts of human genesis that had been used in the seventeenth century. Earlier Christian writers such as Isaac de La Peyrere had argued that whereas Western peoples had descended from Adam, the uncivilized, darker-skinned peoples in "the world newly discovered" had, like the Egyptians, Ethiopians, and Scythians (whose recorded history predated biblical genealogies) descended from a non-Adamic strain.[33] Later, nineteenth-century biblical ethnography assumed all humans to be descended from the original Christian pair, but held that non-Christian savages had degenerated in a separate history.[34] Stocking points out that the hypothesis of physical and cultural degeneration provided anthropologists with a useful explanation for the increasing diversity of "savage" peoples that were being discovered in the early nineteenth century.[35]

The local missionaries were a source of ethnographic information for researchers such as Marshall, in addition to providing food and shelter in regions far from towns, and acting as guides and interpreters of languages they often had a fair knowledge of. Their purpose in being there was closely connected with the assumptions of biblical ethnography. If these races had perhaps descended from a degenerate strain of Adam, or were survivals of ancient non-Christian cultures, here was a vast population that had either forgotten or had never heard the word of God, and must therefore be in need of enlightenment. Further, if Brahminism was the highest form of civilized culture in India, these races were bound to develop toward that form, and then, caught up in the complex philosophy of Hinduism, they would be lost to Christianity forever. This process of Hinduization of the tribals of India, then, needed to be preempted by bringing Christianity to them.

Take, for instance, the following concerns of G. W. Blair, an Irish Presbyterian who in his 1906 memoirs combined the history of his "jungle tribes" mission with ethnographic-style accounts of the tribes he worked with. In the introduction to his book, he quotes Max Müller, Alfred Lyall, and William Hunter to support his concerns that Hinduism is fast claiming the tribes of India. "It would seem almost impossible to exaggerate the importance, from a missionary point of view, of the early evangelization of the aboriginal tribes of India. . . . A gradual assimilation of the aboriginal tribes and their more or less complete transmutation into Hinduism have been going on for centuries and are now in actual process."[36]

Citing Sir William Hunter's prediction that most tribes will be absorbed into "higher faiths" in the next fifty years, when Hunter warned that "it rests in no small measure with Christian England whether they are chiefly incorporated into the native religions or into Christianity," Blair takes up the challenge. "What a reinforcement for the powers of darkness if these tribes pass over into some corrupt form of Hinduism, and how terribly acute must be the subsequent conflict with Christianity! but if—and God grant that it may be so—these caste-less millions accept the faith of Christ, . . . then assuredly the day of India's redemption will speedily dawn, and this land of our love will thereafter become to us dearer still as the abode of righteousness."[37] Although Blair's mission operated in central India, he was articulating an opinion common to all the colonial missions in the country, which drew support from sources as diverse as Indologist, Utilitarian, and anthropological writings.

In south India, one of the most active missions was the London Missionary Society (after the Basel mission, probably the most concerned with tribal conversion). Missionary Samuel Mateer of this society is well enough known for his ethnographic work to merit a mention in L. P. Vidyarthi's 1978 anthropology textbook for his contribution to anthropology. His 1883 *Native Life in Travancore* has detailed sections on the habits and customs of the tribes inhabiting the Western Ghats just south of the Nilgiri range. Describing their mode of shifting cultivation, he remarked disapprovingly that it yields high returns for less labor, and that "it is highly destructive of valuable forest lands." This habit, therefore, he ascribed to their "laziness," recommending that although they preferred "this savage life," they "should be encouraged to settle if possible: only by such means can they be reclaimed to civilisation and education."[38]

Both Marshall (1873) and Mateer (1883) use a narrative in which the capacity to labor is a crucial measure of civilization. In this period, the 1870s and 1880s, the Nilgiris were being rapidly settled by Europeans (as we have seen in chapter 2), and the idea that indigenous tribes had not used the pleasant climate and rich soil of the region to build settlements such as the English were doing was seen as evidence that, in their nonproductive relationship with nature, they were exhibiting their inferior civilizational status. Returning to Marshall to illustrate this point, we can clearly see the link between civilizational progress, nature, and labor. "These hills are covered with good soil—indeed in the moist hollows it is pre-eminently rich and productive, and the land is very accessible to the plough. There is excellent clay for pottery. A laborious, acquisitive race, conserving the glorious water supply, would render this land a paradise. But the Toda scheme is far simpler. He has cattle who afford him all he wants; why should he work? Why should he plough? And from the lazy man's point of view, perhaps he is right."[39]

Once again we see laziness and labor positioned in hierarchical scheme of racial and civilizational progress. Moreover, the incapacity to labor is, for Marshall, linked with an incapacity to appreciate nature as Western races do. Confirming, through head measurements, that the Todas lack the organ of "taste" for beauty, Marshall describes the natural beauties of the Nilgiris for the appreciation of his civilized readers, and goes on to contrast this with the Toda's apparent lack of interest in this beauty. The view from Mukurti Peak, he says, "seems to unite all the possible beauties of such a land. Such a mixture of the lovely properties which a landscape may derive from the presence of stupendous depth, brilliantly transparent and buoyant atmosphere, delicate distances, changing tints, wonderful shadows, I have not seen even in the Himalayas."[40] And, describing the dawn in an elaborate three-page rhapsody: "'Tis the passage of Aurora! She sweeps lightly along over the drooping grass stalks, scattering their burdens as she goes; reminding one of all that is fresh and cool—fountains, crystal, the happy tinkle of silver bells! . . . This wonderful country, ever beautiful and expressive, silent yet speaking, quiet and secluded, forms the beau idéal of a breeding-place and nursery for infant races."[41]

In this description there is a strong sense of an Edenic paradise from which man was expelled, only to degenerate to his present state: once again, signs of a biblical ethnological framework. These beauties, he asserts, the Toda has "leisurely" roamed upon for centuries without appreciating, seeing in them merely opportunities for meeting grazing or fuel needs. The Toda has "omitted all notice of these extraordinary beauties, . . . [as he is] a tasteless man.

. . . He sees the grass. Ha! He sees the dew. Ha! He sees the forest. Ha! But apparently it is only so much cattle's food with water on it, and fuel in the distance."[42] In this representation, the Toda is represented as viewing nature only as a resource, rather than appreciating its "higher" qualities. Do we see a switch here, in which the Toda is now the "user" of nature, and Western man the contemplator? How can we understand the juxtaposition of this model with the previous one of the Toda as the lethargic nonimprover of nature, and civilized man as the user (through labor) of nature?

As in the literature on shikar, we see a romanticist contemplation of nature offered along with an aggressive model of subduing nature to social ends by transforming it through labor. And again as in the narratives of the Ootacamund landscapes, we see the capacity to see the "higher" beauties of nature as a yardstick of civilization. The Todas are seen as failing both tests of superiority. First, they view grass and water as useful natural resources, rather than in terms of fountains, crystal, and silver bells; and the dawn as the beginning of a day rather than as the coming of Aurora—they lack the graces of higher cultures. Second, they fail to transform the nature around them into a productive resource, thus remaining in that primitive stage of civilization that precedes private property and industrious commerce. Along with the attainment of a high culture through the application of labor to nature, according to this model, come many new perceptions of nature, one of the most important being the recognition of a high nature, corresponding to this high culture. A nature-culture opposition is carefully nurtured: the viewing of nature as a pristine space, untouched by crass human needs, is key to the appreciation of this high nature, and can be attained only by men of high culture. As evidence of this, we might note the metaphors used to describe the dew-tipped grass: fountains and crystal, signs of a sophisticated civilization. Low (untransformed, unappreciated) nature goes along with low civilizations. Significantly, the nature-culture opposition is not maintained here; the Todas (and other tribes) are repeatedly described as part of nature, their social and religious beliefs caught up in natural formations.[43] Indeed, these tribes are held to be so much a part of nature that they cannot truly see it. In order to appreciate nature, one had to be outside it; and one traveled outside it only by transforming it through labor. By now this scheme should be familiar, as we have seen it operate in various contexts.[44]

Biblical ethnography exemplified by this ethnographic study of the Todas of the Nilgiris is one piece of evidence of the close ties between missionaries and the visiting ethnographers that they aided; missionaries themselves created ethnographic knowledge of the tribes they worked among. The

Reverend Samuel Mateer's ethnographic work, *Native Life in Travancore*, was linked to other discourses of labor, as well.

Mateer worked mostly among the Kanikars, a tribe whose induction into forest labor was discussed in chapter 3. Mateer complains that "the great vice of these mountain men is drunkenness." He describes a group of London Missionary Society preachers traveling in the hills, and coming upon a group of Kanikars "assembled with jars of arrack," which they were about to offer in worship and which they would later, as was the custom, drink themselves.[45] The missionaries began to preach; but hastily departed when the "headman" reportedly shouted, "Children, make a pile of wood at once to burn these fellows! They are come from the white men to take us to their company and make us eat beef."[46]

Although it took the emboldening influence of spirits to come up with a tactic like this one, the hill tribes were rarely loath to argue with the itinerant preachers who, with a host of anthropologists, foresters, and planters, were becoming familiar sights along formerly quiet forest paths. Mateer relates conversations in which tribals decline induction into Christianity with the explanation that "if they embraced it, the jungle-demons would be offended, and send elephants and other wild beasts to kill them and destroy their cultivations." The sight of bibles in the hand of itinerant preachers would evoke such comments as: "Do you come to destroy us by bringing the wrath of the demons upon us?" One woman said, "I have only two children; do not kill them by teaching them your Vedam [religious book or knowledge]." When asked "why then, do not the Europeans suffer, who cut down the forests?" they are reported to have answered, "As the white men worship a mighty God, the demons take flight."[47]

In the last response, the Kanikars are subscribing to an explanation of European power that was in fact commonly offered to them by missionaries, namely, that the Christian god was mightier than any local deity.[48] The Christian god, in this explanation, oversaw, managed and protected the activities of the colonial foresters. This narrative mixes up claims of spiritual and scientific power; the result is not to weaken or taint the modern by the premodern, however, but rather effectively to strengthen colonial power over the forest and its inhabitants. Casting scientific issues in spiritual terms is not just a rhetorical device so that missionaries may appear to be "speaking the primitives' language"; it is, at one level, a conversation precisely about instrumental efficacy, and at another level, an argument about divine sanction for the worldly appropriation of power. Scientific forestry and Christian

FIGURE 10. Kanikar. From Edgar Thurston and K. Rangachari, *Castes and Tribes of Southern India,* vol. 3 (1909; reprint New Delhi: Asian Educational Services, 1987), facing 177.

proselytizing support each other's claims to power through a mutually sustaining set of discourses and practices. At the level of instrumental rationality, scientific forestry had easily visible, indeed spectacularly obvious, effects on forest landscapes. And missionary claims about the spiritual superiority of their god were buttressed by these effects at the level of local inhabitants' everyday experience.[49]

The missionaries and anthropologists who entered tribals' living areas could do so only because of the sustained pressure on these areas by the forces of scientific forestry, which brought surveyors, mapmakers, roads, communications, deforestation, and replanting activity into the villages and pastures of local groups. Missionaries could assert the superiority of their religion by virtue of the effect that forestry was having on these areas. Without the power of these developments sustaining the presence of missionary groups, they, too, would have been at the mercy of the seasons, the rain patterns, the animals, and the diseases of the mountain forests. It is these environmental factors that the tribals identify as organizing their own subsistence activities and cultural lives, and they correctly identify the "white men" as somehow immune to these factors, or as organized by a different, more powerful, set of factors. As missionaries were fed, clothed, and sheltered by distant funding bodies, protected by a distant government, and welcome in the houses and clubs of the white settlements, it must indeed have seemed a mighty God that they served.

The context provided by Forest Department works and plantation lines, in fact, turned out to be convenient places for missionaries to bring their civilizing messages to tribals. Mateer recounts how, not able to defend themselves against malaria or counter the hill tribes' self-sufficiency in their own surroundings, missionaries made a regular habit of visiting coffee plantations, and thus came to form what were known as cooly missions. A Nilgiri planter, assessing the benefits of cooly missions, suggested to Mateer that

> Coffee planting in this country seems as if largely intended, by the Providence of God, for the good of this hill tribe. In the plantations Kanikars and Christians meet and work together, and some of the latter . . . urge their companions [to convert]. . . . Catechists speak with labourers; spend the night, and next day preach in valleys. These people have already learnt one or two valuable lessons. One is, that the spirits they worship have no power over Christians from Europe and the plains. . . . Another lesson they have learnt is that Christianity is a civilizing and an elevating religion, and a good religion for

> this life generally. . . . The hillmen despise the Pariahs and Puliahs, but they see that our converts have wonderfully improved.[50]

In this planter's narrative, the tribes simultaneously learn of the spiritual power of Christianity and the material benefits of conversion. The former is an assertion on the part of the missionaries; the latter is an effect of the instrumental efficacy of scientific forest management and the wordly power of missionaries as a result of their position within the colonial society and economy. The two statements are not contradictory—rather, they are mutually sustaining.

Economics, science, and religion, thus, were interlinked in a vastly more powerful system of knowledge and practice than the local tribes could muster an effective defense against, either in terms of a contest between systems of knowing nature or in terms of their instrumental efficacy.

That these processes were constitutive of more than just a specifically nineteenth-century version of modernity is clear from the ways in which modes of thought and frameworks for understanding tribal peoples in India retain even today a characteristically moral-scientific tone. L. P. Vidyarthi's text, *The Rise of Anthropology in India*, explaining the changes in Kurumba lifestyles in the 1960s, refers to the publication of a series of studies by the Anthropological Survey of India in 1964–1968. He comments:

> It is fascinating to note how the clearance of forest has brought about considerable change from traditional economy of hunting, fishing and primitive agriculture to improved cultivation of cash crops and wage economy. The Mulu Kurumbas have been attracted to raise ginger and termeric, pepper and coffee which fetch ready cash, and thus integrate the Mulu Kurumbas economy with the wider market economy. Along with their changing economy and wider contacts, their daily routine, their material culture, their concept of cleanliness and sanitation has also undergone changes specially during the last two decades.[51]

The familiar evaluative framework of transitions from traditional to improved, from primitive agriculture to wage labor, and, most interestingly, from an implied state of uncleanliness to clean living, operate here, as in the colonial ethnographies. I have chosen to quote repeatedly from Vidyarthi's text because it is a standard graduate anthropology text, and as such is representative of the body of knowledge that is, even today, transmitted pedagogically in the continuing process of maintaining anthropology's disciplinary status as a science. The ideological constitution of scientific knowledge has remained largely unexamined in postcolonial India, precisely because both scientific communities and lay society assume the conditions of production of such knowledge to be irrelevant to the truth of the theories themselves.

Knowledge produced by colonial anthropologists had effects on tribal populations' lives, organizing their education and restricting their survival options as a result of its refractions through official policy.

Education, Industriousness, and the Moral and Political Significance of Ethnographic Ways of Knowing

Educational and political choices were often made on the basis of the epistemological framework provided by ethnographic ways of knowing. Tracking ethnographic knowledge outside the realm of professional or academic discourse makes evident its implications for the policy decisions that affected the day-to-day choices and restrictions that tribals faced.

W. E. Marshall's phrenologically based correlation between primitivity and lack of taste might seem idiosyncratic, especially as it was phrased in terms of skull protuberances, a diagnostic system now dismissed as pseudoscientific.[52] The assumption that colonized natives lacked taste, however, and the linking of this to a lack of both sophisticated culture and respect for labor, was widespread among scientists and educationists. This scientific way of knowing was influential beyond the domain of professional scientists.

In 1850, a British surgeon, Dr. Alexander Hunter, opened the Madras School of Industrial Arts "entirely at his own charge, with the liberal and enlightened design of creating among the native population a taste for the humanising culture of the fine arts." Among its explicit aims was "to improve the taste of the Native Public, and make them familiar with beauty of form and finish, in the articles daily in their hands and before their eyes."[53]

The significance of combining arts education, cultivating aesthetic sensibilities, with schooling in industrial skill was once again expressed through the category of labor as that which simultaneously created civilizational virtues and met the immediate material needs of the empire. The school's report sees potential in the vast mineral deposits of the presidency, and lays out plans for the development and utilization of these and other natural resources. "[V]ast benefit will assuredly accrue to the people of this country, through the opening up of new and extensive fields for the employment of labour, and the acquisition of wealth, and increasing wealth invariably brings in its train increase of civilization and enlightenment. Nor will the people alone benefit; for such is the framework of the social system in this country, that no increased demand can arise for the products of the soil . . . without *directly* augmenting the revenue of the state, in a degree unknown in most other countries."[54]

Labor, civilization, and imperial profits are linked once again. The management and commercialization of mineral deposits and products of the soil are spoken of in the same framework (that is, a framework of natural resource management) as the development and management of native labor. Natives stand along with the mineral deposits and the soil of the country as elements that are similar in kind: all are untapped resources, sources of potential wealth for the colonial state.

Soon after the School of Industrial Arts was established in Madras city, Hunter received inquiries regarding the potential benefits of industrial training for the "wild tribes" of the presidency. These inquiries, and Hunter's response, were predicated on ethnographic reports of tribal lifestyles such as the examples we have already seen. For instance, a 1862 letter to Hunter reported:

> The Yenadis are a wild tribe. . . . They employ themselves in collecting the products of the jungle which they deliver to the officers of Government receiving from them the value thereof partly in grain and cloth and partly in cash. They are exceedingly illiterate and averse to quitting their hereditary habits which they seem to fear would be attended with evil consequences. . . . An attempt was made to induce these people to adopt agricultural pursuits and this must be pronounced as hitherto a failure. Neither are they disposed to rear cattle, as they have hardly any pasture for them to feed on. In fact they have shown the greatest repugnance to engage in any pursuit they are unaccustomed to. On their condition becoming known to the Home authorities, orders were issued for the adoption of measures likely to effect a gradual amelioration of the state of the tribe; and the establishment of a school for the education of the young was one of the measures sanctioned by the local Government to effect this object.[55]

The civilizational status of the Yenadis here becomes a condition that it is necessary to ameliorate. The certainty associated with this diagnosis and its remedy mark such judgments with the objectivity reserved for scientific pronouncements. The civilizational hierarchy that had been established by the academic discipline of anthropology is the starting point for the observations and judgments made here. The conditions of production of that anthropological knowledge are invisible, the ideological basis for this erased from this knowledge, so that it is available to colonial officials as an objective basis for their procedures and policies.

The writer of this letter, V. Kistnamachari, a deputy inspector of schools, recounts his dissatisfaction with the kind of instruction that the Yenadi boys are receiving in the existing school, which was begun on a government grant in 1858. In 1861, the school reported, "The Yenadis being still in a wild state attach little or no value to education and the attempt to discipline and instruct their indocile children is a laborious work; but I am satisfied that our school is doing some good for them though slowly."[56]

Within the civilizational progression of hunter-gatherers to settled laborers, or wild to docile, the education of the Yenadis can be seen as a moral project. To accede to a state of disciplined docility here sets the parameters of the scheme of moral progress imposed on the Yenadis. Kistnamachari is, moreover, an educationist with a mission to reform what he sees as an inadequately attentive system. He expresses the desire to transform the school curriculum from one of book learning to a more vocationally oriented industrial training school. Keeping in mind that tribals made useful forest guards, he envisions a school that would train them to work with local jungle products such as roots, soap nuts, avarampattu, and rattan. Such training, apart from giving them "some useful calling in life, will not fail to raise them intellectually and morally," he argues. Writing to Hunter for advice, he concludes, "the more I see of the working of the school and the habits of this rude tribe, the stronger is my conviction that . . . it is desirable to modify the character of the instruction imparted in the school as to develope [sic] among the boys habits of useful industry, steady application to work, and even a love of labour."[57]

Hunter sent a favorable reply, and subsequent years saw the Industrial Arts School at Madras become steadily more involved with the ethnographic study of tribes. In 1866 a set of photographs of all the major tribes of the Nilgiris was commissioned by the School.[58] In 1867, wax and bronze casts of the hands and feet of hill tribes were sent to exhibitions in Paris and London, and to other Indian Schools of Art.[59] In 1868, drawings and photos of hill tribes were sent to a Bombay Exhibition.[60] Thus, alongside the development of a system of education that trained tribals to be of service to the Forest Department, there developed a vigorous ethnographic research program that used these same tribals as a source of scientific data. Again, tribals were constituted as an object of scientific analysis by the same processes that developed them as resources in order to sustain the profitability of a system of resource extraction that itself was part of a global economic system.

Although neither Hunter nor Kistnamachari could by any stretch be considered anthropologists, the model of knowledge about tribal lifestyles that they both employed is clearly ethnographic. Ethnographic ways of knowing thus permeated different spheres of understanding and doing, in ways that could not be completely understood if we were to restrict our historical investigations to the sphere of professional or disciplinary anthropology, or formally ethnographic treatises. The important point is that ethnography as a science had, by the second half of the nineteenth century, established itself as a common, indeed indispensable, means of perception of nonindustrial societies. Ethnographic knowledge was, as George Stocking has shown, a product of its geopolitical context, and thus in the nineteenth century incorporated certain ideological assumptions about the relationship of nonindustrial to industrial societies. These assumptions were often cast in moral terms, but packaged as they were within the science of ethnography, they could be presented as objective knowledge of the stages of civilization, and consequently were widely used in prescribing schemes of progress for tribal populations.

A particularly stark example of how the hidden evaluative judgments in ethnographies found their way into social policy is given by the following account of the south Indian famine of 1876. The following is an extract from a local paper that was reprinted in an official report of the famine. The newspaper correspondent, in the course of the article, quotes from popular school geography text.

> [T]he streets were lined with paupers, mostly from the surrounding villages, and many from the jungles, who were waiting for . . . the daily distribution of food. . . . The jungle people are called Yenadies, and are in the habit of coming into towns to sell honey and other jungle produce. The following account of them is from a popular School Geography: "A wild race of people called Yenadies inhabit the jungles along the sea-shore. In habits, religion, and languages, they are quite distinct from their neighbours. They are short in stature and of black complexion, capable of enduring great fatigue, and remarkably faithful and honest. They live on roots, fruits, leaves, fish and rats, and have *no industrial pursuits*." These people have been driven from their haunts in great numbers by the famine, and a large proportion of the paupers in Kalastri belong to the class.

> In accordance with their *idle and unsettled habits*, they will not work nor submit to restraint, and many of them must therefore die of want.[61]

Here ethnographic information on the Yenadis has traveled, by way of a geography text and a popular newspaper report, into the official reports of the state. Offered to the reader as information on the famine, the social implications are nevertheless clear: populations that are strangers to the modern notion of work are of less value to the state. Much of famine relief was carried out through offering food for work, and many roads and water tanks were thus built during famine periods, with cheap or free labor. Tribals, even in ordinary times fearful of urban routines, very likely found the work camp discipline intimidating, and consequently fell outside of most relief programs in many famine periods.

Another effect that ethnographic ways of knowing had on the day-to-day life of tribals may be seen by examining the legal restrictions that were introduced into several tribal populations' lives with the intention of improving law and order in colonial society.

Criminal Tribes and the Practical Applications of Ethnographic Ways of Knowing

Religious moral teaching, imperial interests, and ethnographic knowledge were closely and explicitly linked in the treatment of "criminal types" in colonial India. Anthropometric knowledge was enormously popular among officials of police reformatories, and most such institutions commonly had in their office bookshelves copies of standard European anthropometric texts. There was also a vigorous correspondence between police reformatory officials, missionaries, and amateur anthropometrists, through journals and personal correspondence, on the "nature" of the Indian criminal, who was usually identified as low-caste or tribal, and primitive in his or her morals and work habits.

Consider the following account, from *The Juvenile Criminal in Southern India,* by the superintendent of a Madras reformatory. The chapter on anthropometry begins:

> In India, there is a great field for research in criminal Anthropology and Anthropometry. . . . Pointed heads, flat-roofed skulls and receding foreheads are noted abnormalities, and in regard to the ear, . . .

> malformations are said to be more common among criminals. . . . In short, criminals are supposed to present features more akin to the savage races, and to possess corresponding traces of character. . . . With the aid of a pair of calipers, which every school should possess, the breadth of the head can be measured, as well as the length . . . [and the] cephalic index calculated.[62]

The connection between anthropology and reformatories was thus the assumed inherent link between savagery and criminality. As India was considered rich in stocks of both savages and criminals, this afforded both proof of the hypothesis and the opportunity for the collection of anthropometric data on the physiognomy and physiology of criminal types.

The chapter includes numerous plates of shaven boys, grouped according to caste. One plate shows six pairs of bare feet, and the caption reads: "This plate exhibits some types of feet among juvenile criminals. One chief characteristic is their flat-footedness. . . . The pair of feet third from the right is prehensile in character and closely resembles those of a monkey."[63] Once again, we see the morally charged project of reforming the natives occur simultaneously with the use of tribals as an anthropological data base. Even as the science of anthropology was becoming more widely practiced, it was both constituting and being constituted by the moral and political discourses and practices of the time.

Reformatory uplift programs included religious instruction and the "singing of clean high-toned songs and the narration of interesting tales, with definite morals." Coombes writes that morality and patriotism were instilled in the boys, in order to elevate their character: "Moral training is based on the broad principles of morality common to all religions. . . . The importance of the Sunday lesson cannot be exaggerated, for it is found to be one of the most powerful means of elevating a boy's character. . . . A sense of loyalty to the British Raj is maintained with the aid of the Union Jack, which is hoisted on the anniversary days of the Royal Family, and on other high days. It is explained to the boys that the flag stands 'for justice, good government, and liberty.'"[64]

Criminal anthropometry was by no means a marginal science, in terms of either its community of researchers or the texts it placed itself in relation to. Ethnographic data were used widely in the police departments of the Raj, and were expanded considerably after 1857. Frederick Mullaly of the Madras police, author of *Notes on Criminal Classes of the Madras Presidency*, was in 1893 appointed honorary superintendent of ethnography for Madras

Presidency, a post instituted by Herbert Risley. In his preface, he declares, "These notes on the habits and customs of some of the criminal classes of the Madras Presidency have been collected at the suggestion of Colonel Porteous, Inspector-General of Police, and put in the present form in the hope that they may prove of some value to Police officers who are continually brought in contact with the predatory classes, and of some slight interest to such of the general public who may wish to know something of their less favoured brethren."[65]

After 1893, police officers in Bengal and Madras were required to have certificates of proficiency as anthropometric measurers; anthropologists such as Edgar Thurston trained police in Madras. The inspector-general of police in Madras, Colonel Porteous, wrote in 1894, "[I had earlier] expressed the opinion that the anthropometrical system for the identification of habitual offenders was too Scientific and too dependent on extreme nicety of measurement and mathematical accuracy to be suited for universal adoption in this country; a more practical acquaintance with this subject has led me to modify my views."[66]

E. R. Henry, the inspector-general of police in Bengal, noted, "With anthropometry on a sound basis professional criminals of this type will cease to flourish, as under the rules all persons not identified must be measured, and reference concerning them made to the Central Bureau."[67]

A practical extension of anthropometric data was the typology of various caste "personalities"—thus a particular caste (or tribe) would be classified as "criminal," "martial," "docile," "loyal," and so on. Anthropological studies were drawn upon in the production of various applied anthropology manuals that provided important information for the British administrator at every level. Agricultural trainees employed by the Indian government were required to read two books outside of their agricultural texts—one of which had to be an anthropological work. Caste handbooks were used for instruction of officers in the Indian Army. They contained tables of descent for various castes—Pathans, Gurkhas, Jats, Gujars,and so on—and suggested good recruiting grounds for native soldiers.[68] The Criminal Tribes Act of 1871 made it possible to proclaim entire social groups criminal, on the basis of their ostensibly inherent criminality.[69] Subsequent to the enforcement of the Criminal Tribes Act in Madras Presidency (1911) there was a rash of publication of notes, lectures, and other forms of information on the "criminal tribes of Madras Presidency."

Muhammed Abdul Ghani, a lecturer at the Police Training School in Vellore, in his 1915 *Notes on the Criminal Tribes of the Madras Presidency,* assessed the Yenadis, a tribe we have heard about earlier in this chapter. Ghani describes them as an "aboriginal class" inhabiting the Nallamalai Hills, living in a "state of barbarism," subsisting "chiefly on forest produce such as fruits, roots and game which they hunted at pleasure, until they were enslaved by the Reddis." He quotes Edgar Thurston extensively on ethnographic details of their habits: they knew forest flora well, as also the practical uses of trees and shrubs, herbs, and roots, could make fire with sticks, were expert shikaris and anglers, were addicted to drinking, and so on—a list of their useful and dangerous attributes. "As they remained in the primitive hunting and fishing state without possessing any culture or civilisation," he writes, "endeavours were made by Government to ameliorate their condition." This involved grants for opening schools and the establishment of "industries," the assignment of land for cultivation, and so on. These efforts were found to have "no effect," and he reports the Yenadis to have been "recently settled under the Criminal Tribes Act."[70]

Descriptions of several other tribes, complete with ethnographic details, not only exhibit all the by-now familiar essentializations but also permits a picture to emerge in which tribe after tribe is shown to have lost its livelihood with the advent of various modern systems, such as forest management and the railways. For example, the Donga Yerakalas, who were originally grain carriers who would travel long distances with their pack bullocks, carrying grain from farming areas to markets, had lost their livelihood with the advent of the railways and improved transportation. As we have seen with the Chenchus, nomadic populations were seen by the state as a threat to the safety of settled property owners; wandering tribes without livelihoods could easily subsist by robbery, escaping immediately without trace, as they had no permanent residences. The Criminal Tribes Act was the only means by which to enforce a settled lifestyle upon them. Once nomadic tribes had been made sedentary, they would then be dependent on the state to provide them with means to gain a new livelihood, and were completely under the control of the state. At this stage they were provided with land grants or advances for the purchase of buffaloes and ploughs, which would start them off on the road to becoming agrarian, industrious tribes.

Under the Criminal Tribes Act, tribes or persons believed "to be addicted to the systematic commission of non-bailable offences" could be "notified" as criminal. The act provided for "the settling of criminal tribes in organised agricultural or industrial settlements, . . . [and] for the reformation of the inmates and their introduction to honest methods of gaining their livelihood." Among its provisions were:

— Local Government may vary specified place of residence, . . . [even] into another province.
— Local Government may establish industrial, agricultural, or reformatory schools for children; and may order to be separated and removed from their parents or guardians and to be placed in any such school . . . the children of members of any criminal tribe.
— Whoever, being a registered member of any criminal tribe is found in any place under such circumstances as to satisfy the Court (a) that he was about to commit, or aid in the commission of, theft or robbery, or (b) that he was waiting for an opportunity to commit theft or robbery, shall be punishable with imprisonment and fine.
— Every village headman . . . , watchman, . . . owner or occupier of land on which such persons reside . . . [must report on criminals' movements—failing which he] shall be deemed to have committed an offence punishable under . . . the Indian Penal Code.
— Members of tribe must request (in triplicate) permission to leave settlement; [travel route must be specified], and he shall travel by no other route. He shall report himself to the headman of every village in which he halts overnight, and such headman shall affix his signature on the reverse of the pass . . . with hours of arrival and departure.
— After being discharged, members must maintain fixed residence, attend roll call, obtain passes for travel, maintain good conduct send children to school till they are 12 yrs of age.[71]

The extent to which mobile populations were seen as threats to a stable society is well illustrated by a set of *Lectures on Some Criminal Tribes of India and Religious Mendicants* by C. W. Gayer, principal of a police training school. This text is explicit in its indebtedness to academic anthropology and colonial history, citing among its sources work by Monier Monier-Williams, Alfred Lyall, George Grierson, Herbert Risley, H. H. Wilson, William Crooke, and numerous other well-known Orientalist and Utilitarian scholars of the eighteenth and nineteenth centuries. Accompanying Gayer's lectures, the prescribed texts for the course were listed as E. J. Gunthorpe's *Notes on Criminal Classes,* William Crooke's *Tribes and Castes of the North-West Provinces,* Matthew A. Sherring's *Hindu Tribes and Castes,* F.-F. Mullaly's *Notes on Criminal Classes,* and Henry M. Elliot's *Races of the North West Provinces of India.*

In addition to the regular chapters on various criminal tribes, Gayer includes a long section on the dangerously nomadic populations of religious mendicants. Although it is not commonly recognized, he says, the police play an important part in "social reform" of India, since "certain sections of Indian society sanction and even encourage many practices [against] the law of the land." Religious ascetics who, in accordance with Hindu scriptures, survived off alms from people, were especially threatening because their wandering habits made them impossible to monitor; thus they were an unpredictable element. Further, if they were mobile, they could not form a predictable labor supply: this lack of discipline and work made them clearly an example of an uncivilized and dangerous type. The "incapacity" to labor, as we have seen in other instances, was represented as a dual failure—a moral failure, as work was the means to individual progress; and an economic failure, as loss of labor meant loss of profits for the empire. Both these conclusions are explicitly drawn by Gayer, who cites census data to estimate that there were five million beggars counted, who received in total two million rupees in alms from Indian people; the "loss of their labour and the burden of their maintenance" was thus a "great loss to the Empire."[72]

Gayer speculates at length on the evil inherent in a lifestyle that pays no respect to the virtues of work. In keeping with common ethnographic arguments that saw the backwardness of contemporary primitives prefigured in medieval European habits, he compares India to "Roman Catholic" Europe, which in "medieval times" was "overrun by religious mendicants" who, under the strain of a secluded life, "gradually fell into slothfulness." The cure for this European malaise, he argued, had been "civilisation, enlightenment, industrial development, and conscription," of which he judged the first three, if successfully introduced to India by the colonial state, could eliminate the dangerous habit of mendicancy. Emphasizing how much criminality was tied up with religion, he used as an example the sage Valmiki, author of the religious epic *The Ramayana*: "Perhaps the best known instance of a man who would nowadays certainly fall into police hands, becoming a religious recluse is that of one of the most famous robber chiefs that ever lived, Valmiki."[73]

Giving examples of the political danger posed by religious mendicants, he reports: "Quite recently some so-called Hindu Missionary Societies have taken advantage of the Govt's gift of freedom and have organized seditious propaganda and are employing Sadhus to spread sedition. . . . [A] certain Hindu missionary body in Calcutta . . . is now occupied in preaching active hostility to the British Raj. . . . [T]he police have a suspicion that a large number of the so-called Swamis and Sadhus now touring in the disaffected tracts are there rather to embitter the lower orders than to turn them into paths of peace."[74]

He shows, however, how a knowledge of religious and cultural habits of the natives can aid in arresting these dangers. What follows is a list that looks much like the section headings of the standard ethnographic treatises of the time, which he offers as guidelines according to which he invites information from officers and students. The list includes such categories as:

a. The real home of the gang and its mother tongue
b. Its composition
c. The places visited by it and the dates and duration of visits, and the places chosen for camping and mode of conveying equipage
d. Offenses
e. Disguises
f. Religious beliefs . . . tribal gods, shrines most sacred.
g. Miscellaneous interesting facts such as omens observed, slang expressions used, etc.[75]

His descriptions of criminal tribes, in his lectures, usually included the marriage rituals, social customs, legends, language and common phrases, castes or subsects, markings or tattoos, clothing and appearance, and so on, and read exactly like ethnographic descriptions.

It is worth noting that even as this body of scientific data about local tribes was being established, these tribes were losing their own sciences. Many so-called criminal tribes made use of a practical science that facilitated their activities. For example, the Patharries, a tribe living in the Central Provinces, reportedly made fake gold ornaments from brass by mixing into molten metal the bark of various trees (*Ficus religiosa*, *Tamarindicus indica*, *Ficus glomerate*, and *Bassia latifolia*). A medical officer's report notes that a "gold-like sheen" was produced by the "volatilizing of sulphur and resin." (The same officer goes on to report that "cheating and swindling are the chief criminal pursuits of this caste.")[76] The Chapparbands, Muslims from Bombay Presidency, reportedly made counterfeit coins out of an alloy of tin and pewter. They made their own ovens and molds, and developed specific milling processes. A group listed as "professional poisoners," and counted as a subsect of the "Thags" under the Penal Code, reportedly put dhatura seed in victims' food and hookahs.[77] All these practices might well be considered as forms of technical expertise. Moreover, these were sciences that worked, as they had successfully identified predictable chemical characteristics of organic substances. Ironically, ethnography was a science largely based on erroneous assumptions about race and inheritance. Nevertheless, ethnography as a science was more powerful in this period than the chemical science of local tribes. The two forms of science were differentially imbricated with the political and moral economy of colonial rule, and hence the efficacy of one was curtailed while the other became increasingly powerful.

Although the Criminal Tribes Act was revised and updated for more effective surveillance of habitual offenders in 1924, officials in Madras believed the criminal tribes to remain a tiresome and dangerous problem. As we saw in the case of forestry, the problem of disciplining unruly tribes was proposed to be transferred out of government hands and entrusted to the care of missionaries. For the early twentieth century, this was an unusual step, for Utilitarian policies had by then made it difficult for government to sponsor any religious proselytization. Thus, the police training school lecturer Mohammed Abdul Ghani had, in 1915, tentatively suggested that the state might "come to terms with the London Mission Society, as the removal of the children from the parents appears to be the only means of effecting any permanent change in the habits and customs of the tribe. The London Missionary Society would include moral instruction in the curriculum, but would not attempt to baptise." He was referring to the Koravas, whose criminal habits he said had not been "eradicated" by projects undertaken under the Criminal Tribes Act, which had included their employment in mines (care being taken to keep them separate from other miners, lest their criminal habits should spread).[78]

In 1925, a government note reported difficulties in providing for settled criminal tribes, and announced a proposal to move their management out of government public servants' hands, as this was proving to be a "financial and practical drain." Though it declared itself open to any "non-official agency, missionary or otherwise," including "Hindu or Muhammedan agencies in [the] field of social reform," the government of Madras admitted to having received interest only from Christian organizations. At this time there were six settlements under nongovernment agencies, including the Salvation Army and the American Baptist Mission. Although these settlements were run by these agencies, the government set guidelines based on its long experience with tribal sedentarization projects. Recommending that "there should be some professed religious or philanthropic agency attached to each settlement if not actually in charge of it," they prescribed for these settlements "a combination of agricultural and industrial settlements, . . . the agricultural settlement being held up as the reward for good behavior in the industrial settlement." On the labor question, they recommended employment in quarries, mines, estates, and factories.[79]

This system was fairly successful in providing factory and estate labor in the areas around criminal tribe settlements. Conversely, areas in which tribal labor was considered overly unruly could be declared criminal settlements as a means of disciplining the labor, as under the new proposal a factory or company could count as a nongovernmental agency and could be entrusted with the "management" of tribals. For example, a sugar factory belonging to the East India Distilleries and Sugar Factories Ltd. was in 1919 declared to be a criminal settlement under the Criminal Tribes Act, and the local factory agent made manager of the settlements. In Tinnevelly District, where the Uppu Koravas had earlier been encouraged to work in a local sugar factory, they were declared a criminal tribe, on the basis that they lacked discipline. The factory management set up a cooperative store and dispensary "in order to discourage the Koravas from going out to the village, and established a factory school for Korava boys, who, after graduating, were employed in the workshops and motor depts."[80] The United Planter's Association of South India (UPASI) recruited from criminal tribe settlements for their Annamalai estates, as did the British confectionery firm Parry and Company for their sugar mills.

While employers such as UPASI and Parry were by no means conversant with contemporary anthropological theory, they enter, willy-nilly, into the history of colonial anthropology by virtue of the irreducibly social character of science and the scientistic ordering of society in the nineteenth century. This is not to say that either science or society was fully determined by the other, but rather to assert, again, that a history of science that restricts itself to academic or disciplinary definitions of science will give us only a partial understanding of the workings of science in a particular historical context.

We have seen how moral and political arguments and ethnographic ways of knowing functioned in a mutually constitutive interaction, how colonized tribal natives were constructed as objects of scientific knowledge through the same processes that sought to civilize and educate them, and how the category of labor functioned as a boundary term between the areas of human and natural resources, so that tribals could be constituted and used just as tropical natural resources were harnessed, in order to sustain a global imperial economy. The next section examines the larger ideological context in which ethnographic ways of knowing were of conceptual and practical significance.

Ethnographic Science and Imperial Politics

The naturalizing of cultural difference was assisted by the ethnological treatises that proliferated in the second half of the nineteenth century. Ethnography was no isolated, esoteric realm of scientific inquiry; the conditions under which colonial anthropology was practiced were produced by colonial administrations, while anthropological knowledge, in turn, made possible systems of administrative control.[81]

The objectivist, systematic scientific observer was a powerful constituent of the colonial order. By positioning himself behind the scientific gaze, the creator of official knowledge about the natives rendered himself invisible; the knowledge thus produced was therefore held to be unquestionably true, as the masking of its conditions of production created the illusion that it was interest-free and value-neutral.

Ethnographers studied indigenous people, categorizing them on the basis of religious and caste groupings, tribes, or national boundaries. By treating cultural habits and political formations as inherent characteristics of communities or nations, they hypostatized them, or made them into eternal, static, hereditary characteristics of people and of nations. Once cultural "traits" had been hypostatized into scientific facts, they were available as a resource to anyone who sought to write about the history or culture of the people concerned. Whereas political considerations often operated explicitly in the formative stages of scientific knowledge, once ethnological stereotypes were reported in official tracts, supported by masses of data, they passed into the apolitical realm of incontrovertible scientific fact. As evidence of their conditions of production was erased, they passed into the domain of cultural belief. This is what made it possible, for example, for James Mill, when writing the *History of British India*, to assert that "under the glossing exterior of the Hindu lies a general disposition to deceit and perfidy."[82] Mill asserted that he was able to discern this fact through the power of reason, and not through "direct observation," and indeed had never visited India. It was thanks to an existing, scientifically authorized "way of knowing" nonindustrial societies that Mill had automatic access to evaluative positions that could be presented as objective facts.

The fact that scientific discourse derived its power from an observer-independent cloak of invisibility has been demonstrated repeatedly by historians of science.[83] Invisibility is only half the story, however. Although much of the instrumental efficacy of scientific knowledge did stem from the fact that its claims to objectivity were underpinned by its unacknowledged, invisible links to power, we must remember that science was also often spectacularly deployed, deliberately displaying and acknowledging its links with power in order to increase its authority. Both of these phenomena—the masking of the disciplining hand of the state by a putatively interest-free form of knowledge, and the spectacular exhibition of authority through public displays of power—achieved their fullest expression during the nineteenth century. Nineteenth-century science might seem paradoxical in its mixture of criteria; on the one hand, objective data and unprejudiced observation were supposed to be essential components of good science, which was seen, ultimately, as the disinterested pursuit of pure knowledge. On the other hand, science was seen to embody all that was good and progressive—not neutral, but positive. Science as a progressive, rationalizing force was (most notably to the Utilitarians) the desirable basis of an ideal society.

Literary critic Satya Mohanty refers to the "complementary thematic of invisibility and spectacularisation" as a defining characteristic of racialized discourse in colonial India.[84] What made this "complementary thematic" possible was, I suggest, precisely the growing cultural authority of science. The public power of racial discourse was derived from the ostensible value-neutrality of scientific theories; objectivity and invisibility thus went hand in hand; scientific practices seemed to guarantee the production of decontextualized, eternal truths; the awe-inspiring public displays of authority and scientific efficacy established the practical benefits of scientific knowledge. While the high culture of science was created and preserved as the domain of the intellectual elite, its low culture, or popular face, was designed for and presented to an awe-struck middle and working class, and offered to them as a form of power supposedly available to all. Not only did museums and exhibitions cater to the upper classes but they also incorporated shows and exhibits designed specifically for a working-class audience.[85] New scientific disciplines emerged; new modes of presentation of these knowledges were devised. As Tony Bennett notes, "The space of representation constituted by the exhibitionary complex was shaped by the relations between an array of new disciplines: history, art history, archeology, geology, biology and anthropology."[86] Through these "new" sciences, cultures could be constituted through the retelling of their histories. Each country would take its place in the hierarchy of civilizations, racial types were located in the hierarchy of "man," and everything in indigenous literatures and religions would be given its place in the grand narrative of the history of universal Man. These new sciences affected not only the spectacular, exhibitionary aspects of science itself but also influenced the self-understanding of the social sciences, particularly history.

No work on India more dramatically exemplifies the influence of science on the writing of history than James Mill's *History of British India,* which served as the authoritative text for all civil servants of the colonial administration. Mill's preface details the scientific basis of historical writing, which serves as the rationale for his having written the history of a region about which he had no firsthand knowledge. He explains that the chaos and confusion of India confound the already unreliable senses, whereas the calm, rational intellectual disposition can only be cultivated in the midst of a more advanced culture. He writes that "the powers of combination, discrimination, classification, judgement, comparison, weighing, inferring, inducting, philisophising" are, without doubt, "more likely to be acquired in Europe, than in India." His work, then, purports to assess the truth about Indian history, which, as narrated by natives of India, is full of error and exaggeration: "Rude nations seem to derive a peculiar gratification from pretensions to a remote antiquity. As a boastful and turgid vanity distinguishes remarkably the oriental nations they have in most cases carried their claims extravagantly high."[87]

Mill places Indian civilization in its rightful place in a hierarchy of civilizations, and assesses the practical, philosophical, and spiritual virtues of its customs, crafts, religion (in the singular), and government as they have existed (eternal and static) for over two thousand years. On Hindu myths and legends, for example, he comments: "The offspring of a wild and ungoverned imagination, they mark the state of a rude and credulous people, whom the marvellous delights; who cannot estimate the use of a record of past events; and whose imagination the real occurrences of life are too familiar to engage."[88] The historical narrative thus clearly assumes a certain scientifically legitimated hierarchy of civilizations, with the "wild" or "rude" nations that possess neither a culture nor a history, at the bottom, and the colonizing, globally powerful nations at the apex.

The repeated positioning of cultures in the relationship of primitive to advanced, and the corresponding dichotomous oppositions between sensual and rational, credulous and skeptical, static and dynamic, are not merely rhetorical coincidences but reflect patterns of thought and writing that recent literary theory has exposed in a wide spectrum of nineteenth-century texts. The belief that these qualities (sensuality, rationality, and so on) described features that were inherent in certain populations was grounded in racial theories. This is not to say that every writer who employed these oppositions was conversant with the latest anthropological theories, but rather that "racialized" discourse was part of a set of cultural practices and beliefs that were informed, often indirectly or implicitly, by contemporary scientific thought on race.[89]

That one did not have to be crudely racist in order to have internalized such patterns of reasoning becomes clear when we realize that intellectuals across the political spectrum invariably lapsed into the same patterns of representation of nonindustrialized peoples. For example, Marx and Engels, although their political agenda could not have been further removed from that of Mill, frequently resorted to racialized descriptions of national groups, especially of Irish and of Indians—both at the time under British colonial rule.

Consider, for example, Engels's detailed study, *The Condition of the Working Class in England.* The chapter on Irish immigration refers to the Irish in unmistakably ethnographic style: "Whenever a district is distinguished for especial filth and especial ruinousness, the explorer may safely count upon meeting chiefly those Celtic faces which one recognises at the first glance as different from the physiognomy of the native and the singing, aspirate brogue which the true Irishman never loses."[90] The reference to physiognomy reflects the scientific belief that racial origins and psychological character could be read from the faces of individuals. Franz Joseph Gall's science of phrenology (popularized by William Coombe) and James Hunt's and Paul Broca's development of physical anthropology were influential beyond the boundaries of scientific communities.[91]

Marx wrote regularly on events in India, in the *New York Tribune*. Even as he deplored the exploitation that British colonialism engendered, he saw British rule in India as performing a historically progressive function. Thus he wrote that "England has to fulfill a double mission in India: one destructive, the other regenerating—the annihilation of old Asiatic society, and the laying of the material foundations of Western society in Asia."[92] Marx's vision of Indian history owes much to Mill's account. He sees it as devoid of order, being one long saga of domination and conquest. Indians were supposed to have been destined for subordination. "A country not only divided between Mohammedan and Hindu, but between tribe and tribe, between caste and caste; a society whose framework was based on a sort of equilibrium, resulting from a general repulsion and a constitutional exclusiveness between all its members. Such a country and such a society, were they not the predestined prey of conquest?"[93]

It was a common assertion that colonized countries had no inherent "national" character, had always been fragmented and chaotic, and thus required the external imposition of order. Therefore, Marx holds, hypocritical and exploitative as British colonialism was, it nonetheless performed a function—it dragged Asiatic societies into the bourgeois period of development. He looks forward to the time when "a great social revolution shall have mastered the results of the bourgeois epoch, the market of the world and the modern powers of production, and subjected them to the common control of the most advanced peoples."[94] Finally, he expresses hope in the emerging Indian bourgeoisie: "From the Indian natives, reluctantly and sparingly educated at Calcutta, under English superintendence, a fresh class is springing up, endowed with the requirements for government and imbued with European science."[95]

Historian of technology Daniel Headrick addresses the reasons for the "failure" of Marx's prediction that a newly vigorous, scientifically and technologically proficient class of experts would grow out of the colonial experience, and carry India into a progressive stage of government and social organization.[96] Headrick argues that this failed to happen because of colonizers' vested political interests, which prevented a successful transfer of the "culture of technology" along with the geographical relocation of technologies.

The cultural and technical aspects of scientific discourse and practice are not easily separable, however, and the analytical separation of these aspects leads to a historiography that is at best partial and incomplete with respect to the understanding of both science and colonial societies. If we examine the processes by which both everyday practices and epistemological frameworks were consolidated in particular historical contexts, we can provide better explanations for the failure of Marx's prediction. More important, we can understand the contradictions that characterize colonial modernity not as resulting from colonial indifference to the spread of scientific and technological culture or from the natives' inherent inability to comprehend sciences, but as resulting from certain structurally necessary "functional incoherencies" in the constitution of modernity in nineteenth- and twentieth-century India.

The class Marx refers to in the passage cited above does, in fact, provide an interesting case for us. In attempting to understand how racialized discourse operated, it is useful to trace not only the development of the dominant representations but also the response of colonized groups to this discourse of racialization. Within Indian nationalist discourses, we see an insistent inversion of colonialist hierarchies. Where the colonizing rhetoric constructed them as weak, sentimental, or lost in contemplation of other worlds, the nationalists invoked their own spirituality and gentleness as their unique strengths. The inversion, however, failed to problematize the language in which these beliefs were articulated.

A Nationalist Science: Indigenous Anthropologists

In an earlier section, extracts from a present-day anthropology text allowed us to see the ways in which nineteenth-century assumptions have continued to define the terms in which tribal populations are understood. Although Vidyarthi's textbook applauds British anthropologists, missionaries, and administrators for originating the scientific study of man, its greatest enthusiasm is reserved for descriptions of Indian anthropologists, whose work it describes in glowing superlatives. L. K. Ananthakrishna Iyer and Sarat Chandra Roy are eulogized as the fathers of Indian anthropology, for their work on south and north India, respectively.

Ananthakrishna Iyer held several administrative posts in the ethnographic survey offices of Cochin and Mysore, and produced several-volume works on the tribes of those states between 1906 and 1932. He also held teaching posts in Cochin, Mysore, and Calcutta, and in the course of his career made professional acquaintances with British anthropologists James G. Frazer, Edward B. Tylor, William H. R. Rivers, and many others. His son, L. A. Krishna Iyer, also became a well-known anthropologist, publishing a three-volume work on Travancore tribes and castes and a comprehensive ethnographic social history of Kerala, organized on the basis of Kerala's "pre- Dravidian," "Dravidian," and "Aryan," and other (Muslim, Jewish and Christian) "elements," and including a chapter on serological analysis of "the Negrito strain" of the hill tribes.[97]

In 1942, the Indian journal *Science and Culture* published an obituary for the Bengali anthropologist Ramaprasad Chanda, referred to as "the father of Indian physical anthropology." In a review of Chanda's work, the journal reported that he was influenced most strongly by Herbert Risley and the Italian anthropologist Montegazza. Chanda probably best represents both the nationalist opposition to British anthropology and the extent to which nationalist anthropology was undertaken completely within the terms set by colonial anthropology. The review article remembers Chanda for having challenged Risley's grand theory about Indian racial origins, for having thus "liberated the young science of physical anthropology from the leading strings of officialdom and set it up as an Indian Science which may be pursued by the people of India for their own good, and for unravelling the threads of their own racial history and culture."[98] As Chanda's obituary tells it, much resentment was aroused among the Bengali elite when Risley suggested that, according to anthropometrical measurements carried out by the Ethnographic Survey of India, "there was very little Aryan blood in Indians." Ramaprasad Chanda, it claims, was "one of the few who tried to controvert [Risley's] views on *scientific lines*."[99] He rented anthropometric instruments from the Ethnographic Survey and combined the measurements he arrived at with "information" from ancient Hindu literary and religious texts, engaging several pandits at his own expense to read Panini and "old *Smriti* literature." Chanda argued that the "strong ethnic element" found in the physiologies of the people of the Northwest, Gujarat, the Deccan, Orissa, and Bengal was the result of mixing between Aryans and a "Pamirian race" (who were "round- or broad-headed people"), as opposed to the Aryo-Dravidian, Scytho-Dravidian, and Mongolo-Dravidian mixtures Risley had postulated. These brachycephalic Pamirians were "represented by the Galchas or hill Tajiks of the Pamirs, by the Tajiks of Turkestan, Khurasan and Afghanistan," and were "akin to the Armenoid race of Western Asia and the Alpine race of Europe."[100]

The social resonances of such a correction are easy to see, if we compare Chanda's picture with Risley's. Chanda's racial addition to Indian stock, the brachycephalic Homo Alpinus, was described by him as "A white-rosy race, very brachycephalic, stature above the average, with thin prominent nose, varying from aquiline to straight, long, oval face, hair brown, usually dark, always abundant and wavy, eyes medium." He explains, "In India [this] type has been turned into the mesaticephalic Indo-Aryan of the Outland by Nisada, Vedic Arya, and Dravida admixture."[101]

This racial type was Chanda's correction for the large role played in Risley's theory by the Dravidian race, the descendants of the Vedic *dasyus*, described by Risley as being flat-nosed, dolicho-cephalic, of low stature, thick-set, with very dark complexion, relatively broad face, and low facial angle.[102] Arguing that the Dravidians were actually a mixed race, Chanda said that Risley's description of the Dravidians or dasyus actually referred to the Nisada (pre-Dravidian) type, represented only by the most savage tribes of central and southern India. Chanda protested: "Sir Herbert Risley classifies these dark, short, and broad-nosed savage tribes of Central and Southern India together with the civilized speakers of Dravidian languages under the head "Dravidian type."[103] Chanda argued that the broad-nosed tribals should be classified in a "lower" group than other central and southern Indian types.

Chanda's theory, and the enthusiastic reception it received from Indian anthropologists, suggests the strongly upper-caste Hindu interest in preserving the idea of a "higher" racial origin in Indians. Although Chanda manages to save most Indians from the "dasyu" label Risley wanted to stick on them, it is important to note which Indians are conceded to the other side. These are "twenty-seven broad-nosed jungle tribes with average nasal indices above 84," which Chanda could not justify with classifying with "the upper group," the civilized Dravidians with nasal indices below 80. These most "savage," or "most broad-nosed" of Indians, precisely those tribes whom we see most often in the ethnographic records of missionaries and in Forest Department settlements, in plantation labor, and in criminal tribe settlements, are renamed "Nisada" by Chanda, who traces this classification to the *Mahabharata*, which describes the original Nisada as "a short-limbed person . . . resembling a charred brand, with blood-red eyes and black hair. . . . From him have sprung the Nisadas, viz., those wicked tribes that have the hills and forests for their abode."[104]

This description, Chanda argued, fitted the Bhils and the Gonds of the Vindhyas, and in the Western Ghats it included the Kadirs, the Kurumbas, the Sholagas, the Irulas, the Mala Vedars, and the Kanikars. These are, of course, the same tribes studied for their primitive traits both by Indian anthropologists such as Ananthakrishna Iyer and by missionaries, foresters, and the stream of amateur and professional ethnographers who studied south Indian natives.

A more comprehensive study of Indian anthropologists would be able to push much further the connections between caste status, scientific knowledge, and the broader socioeconomic trends in mid- to late-twentieth-century India. As this is beyond the scope of this chapter, I will close this section with a brief look at an obscure but fascinating work written by an amateur anthropologist, S. Sundaresa Iyer, entitled *How to Evolve a White Race*. This book, in spite of its marginality to academic science, startlingly reveals common perceptions of race and science in the first half of the twentieth century.

Iyer parades before the reader his knowledge of physics, mathematics, biology, anthropology, and mathematical genetics, lest the reader might take him for a mere layman.[105] The book is an elaborate do-it-yourself manual for a nation. India as a nation is globally misunderstood because of its variation in skin colors, but this can be overcome, he announces. In a passionate Lamarckian manifesto, he lays out an elaborate scheme, replete with genetic equations, by which "the normal colour of every Indian can be white with proper knowledge of the Environment."[106] Nature can be overcome with a true knowledge of science, he consoles his Indian readers. It is only an accident of the environment that some races are white and others brown or black, and that white nations discriminate against others on the assumption of racial superiority; this can be easily overcome if, over time, Indians learn how to control all the aspects of the environment their children grow up in, such as "physical conditions like physiographic nature of the country, climate, food supply, mental conditions as Religion, Morality, Legal and Social ideas, marriage and man's attitude towards women, social and economic position, love of the child and of the race, knowledge of the sciences."[107]

In Iyer's work, the most utopian promises of science combine with deeply held caste- and class-based assumptions about privilege and color, to give us this quite disturbing treatise, simultaneously scientistic and brahminical. In its introductory chapter, the "superiority" of whiteness is naturalized with an array of examples—pregnant mothers longing for fair babies, the selection of marriage partners, the varna system, and epic heroes and heroines—arranged so as to convince the reader that "there can be no two opinions about the invaluable nature of white or fair skin in man."[108] The chapter concludes on a note of confidence, when, after having described "the conquest of Nature" through the discovery of its laws, he asserts, "With the help of knowledge handed down to us by great savants in the field of science, we shall see how in the control of skin colour we are on as sure a ground as in the case of steam power, wireless, electricity, medicine, etc."[109]

Such an obsession with whiteness predates colonial rule, and would have to be traced through diverse elements in the religious and cultural history of the Indian subcontinent itself. Nevertheless, the interesting point about it for our purposes is the way it preserves some of the cultural manifestations of science in the historical context in which it was written: in it we can see a preoccupation with the workings of a global relationship between imperial and colonized countries, and a correspondingly global visibility of scientific knowledge and the efficacy of technology. The author's religious and cultural beliefs are refracted through these in particularly scientistic ways, to give us a theory shot through with religious and cultural prejudice, yet presented authoritatively, in the best objectivist fashion of the day. This is, again, a telling example of the mixed status of modernity that I have referred to in previous chapters, and that I will discuss in the context of religion in the following chapter.

Discontents

Finally, I would like to return to where this chapter's second section ended. We saw that the early twentieth century was seeing a more and more successful alliance between the law, mission organizations, and industrial and agricultural employers in erecting a disciplinary and "civilizing" apparatus by which to control a population that was more and more at odds with the general direction of national developments toward organized technological and industrial growth.

Now and then a report indicates breaks in this carefully erected system of discipline. For instance, the 1925 *Note on the Settlement of Criminal Tribes* regretfully concedes defeat in the case of the Koravas, who, although they have been attending the mission school, seem to be getting their own way.

> This [attending school] is not, however, on account of their thirst for knowledge, but so as to be able to prove to courts that they are respectable people when the Police place them before magistrates. . . . The American Mission of Amppukottai at one time had great hopes of converting these Koravas and leading them to an honest livelihood. They converted them, but had themselves to admit that the Koravas had openly embraced Christianity in order to have someone respectable to stand as surety for them when they were run in under the security sections. The mission admitted that the Koravas had not abandoned their criminal habits.[110]

This report is particularly revealing in that it allows us to read the Koravas' intentions, which are clearly at odds with the civilizing mission's. Although occasional reports like this one do not amount to what one could legitimately term concerted resistance to colonial control, it certainly allows us to see for a moment that it was not necessarily the case that populations acquiesced to state and nongovernment attempts to sedentarize and discipline them into a predictable labor force.

There is scattered evidence that several Nilgiris tribes, too, strongly resented the imposition of a new order upon them, although they were often powerless to mount a defense to it, partly because of the multifaceted and complex ways in which this new order was introduced over a period of several decades.

Reverend Friedrich Metz of the Basel Mission Society collected several proverbs and folk songs of the Nilgiri tribes, among which are recorded the following proverbs: "If you appeal to the magistrate, you might as well poison your opponent's food"; and "Riches acquired by serving the sircar [government] are like a post in a swamp, rats carry it away."[111] A funeral song describes people without sin: "They never complained to the Sircar's dubash, / Nor slandered friend or foe . . ." The chorus is a litany of sins, such as: "He killed the crawling snake, the harmless frog" and: "He called in the Sircar's aid / It is a sin."[112]

Expressions of contempt for the government are explicitly recorded in common cultural forms such as proverbs and songs, warning the community of the dangers of having dealings with a government that does not have its best interests at heart. Funeral songs are full of what one might consider "environmental" knowledge, including reminders to respect and take care of the surrounding world. Included after injunctions to care for frogs and snakes, here, is a reminder not to call for government aid in disputes. Disputes between or within tribes invariably brought harsh correctives from the state or other agencies, and it was judged more sensible to leave them out of local matters.

In a song recorded by French anthropologist M. B. Emeneau, the Todas sing about the change for the worse in their living conditions. This complex song seems to illustrate the extent to which the British reorganized intertribal relations, something that is evident from survey, revenue, and census reports. Anxious to determine the exact relations between native populations, survey officials often rigidified loosely structured relations or nonhierarchical interactive networks of local groups when they made official reports; these then became recorded as fact in future reports, on which land revenue, grazing concessions, and so on would be decided. In this way tribal relations would often be altered to a considerable extent over the years. The appointment of a headman, too, was common, in order for the authorities to have a clearly identified individual with whom to conduct official affairs. This post considerably modified intratribe relations, displacing the more dispersed authority structure that had been common.

In the song that Emeneau labels "Song about strange changes from olden times," the Todas describe how the various tribes and individuals were labeled as agrarian, musical, sorcerous, or official.

> They are saying: Times have changed for the worse.
> They are saying: The country has changed for the worse. . . .
> They are saying: The gun is for Europeans in the courts.
> They are saying: The monegar [chief/headmanship] is for the black-headed [Toda].
> They are saying: The fields with crops are for the Badaga men.
> They are saying: The clarinet is for the [Kota] who tied his hair at the back of his head. . . .
> They are saying: Murder is for the black [Kurumba] sorcerer.
> They are saying: Poison is for the black poisonous snake.
> They are saying: Change has increased.
> The horn that was raised high is becoming low,
> The horn that was low is rising high.
> Tell them to say: Tell all the sacred places!
> Tell them to say: Speak to all the gods.[113]

Another song from Emeneau's collection records the helplessness the Todas felt in the face of the disciplinary mechanisms of the state. Sentenced to a jail in Coimbatore, the district headquarters, for stealing a buffalo from the Badaga tribe, two Todas wrote this song while imprisoned. The jail, down in the hot plains, took them far from their familiar grassy hills and into an environment policed by Indians they refer to as "the Tamilians," a group separate from and alien to their own.

> At Coimbatore we have sat in elephant's leg chains.
> At Coimbatore we have been trapped in confinement. . . .
> We have sat longing. We have sat thirsty.
> We are longing for puffed millet in a brass food vessel. We are thirsty for milk in a brass vessel. . . .
> We are longing for the ways that we walked. We are thirsting for the paths that we went.
> We have worn a garment that does not cover the back. We have eaten food that does not reach to the heart.
> We have had the sun hot on the head. We have had the dust burning on the feet.
> We do not look at the crowd (of Todas). . . . We have looked at the midst of the Tamilians.[114]

These songs, moving from the ways in which a strange new government entered the lives of the hill tribes to the newly acquired power of the plains dwellers over the hill people, draw a picture that shows how the new order was not merely a simple hierarchy in which a foreign power ruled all Indians but was also a system into which different "native" groups were differentially integrated. A history of colonial anthropology in all its social dimensions tells us clearly that in this period the tribal populations of south India were inscribed, perhaps more firmly than at any other point in their history, in an extractive economy that involved the intrusion of a disciplining and controlling external force at the level of their everyday practice.

Chapter 6

Christianity

Take, for instance, the work [the British] have done in the coffee plantations of the Wynaad. How much time did it take before those plantations yielded fruit! How many working people had to be engaged in order to prepare the soil and how much money had to be spent! But finally the jungle was converted into beautiful gardens, and these yield now a precious harvest. In the same way Christianity has first to prepare the ground in this land, and to remove the obstacles, as for instance, sorcery, astrology, and others. These must be removed just as the jungle weeds and old roots had to be removed, before the new plantations could grow.[1]

This administrator's choice of metaphor in 1912, linking plantation agriculture (the taming of the jungle and its transformation into productive resource) with religious work (the retrieval and development of lost and corrupted souls) was no mere discursive accident. By the late nineteenth century, Christian missionaries in many European colonies had intricately woven together metaphysical and material goals, increasingly defining their civilizational task in terms of the imperatives of scientific progress. As we have seen in chapters 2 though 5, the management of nature and of natives were theoretically and practically conjoined. Following specific refractions of the discourses around natives, property, and nature through different kinds of material practices, chapters 3, 4, and 5 tracked a dialectical and reflexive relationship between scientific discourse and socioeconomic practice. Analyzing science as practice places it in the arena of agonistic, ideological struggle, alongside other colonial and indigenous moral, religious, and practical ideologies. We now take up a theme that has run through the preceding chapters: that of the moralizing rhetoric and proselytizing activity that linked the discourses and practices of Christianity with narratives of nature and scientific progress in late-nineteenth- and early-twentieth-century south India.

This chapter and the next are linked to preceding ones by an attention to discourses and practices of managing nature, natives, and modernity, but they introduce an additional element. In the introduction, I alluded to global networks both within and across different imperial formations, which become evident when we follow the trajectories of science, technology, or commodities in the nineteenth century. Because of the nature of disciplinary training and research, it is rare for colonial historiography to write its narratives across these European imperial boundaries. Limited by some of these same disciplinary constraints, this book does not attempt anything as vast in scope as the global histories of Eric Wolf or Immanuel Wallerstein. With a more limited palette, however, this chapter and the next include in their purview some actors and resources outside the conventional domain of British colonial history, including Swiss missionaries, Peruvian-Indian botanists, and medicinal alkaloids of the Andes.

The Colonial Missionary's Burden: Progress, Civilization, Morality

Rationalist, technoscientific modernity is often regarded as radically disjunct from the morally charged universe of religion. If we ask how religious and humanist principles were translated into practice, what changes they required from colonized groups, and what specific economic needs motivated the systematization of particular ways of knowing and controlling, we find that religion and science appear contradictory only at the level of official, or high discourse. If we look at lower order or ground-level practices, we can see that this discursive contradiction is really a functional constituent of the kind of order that colonized societies had to be brought into as a result of their structural position in a global network of extraction, production, and distribution of resources.

The assumption that secular scientific knowledge is primarily a category independent of religion takes for granted a separation where often little distinction exists in practice. It is precisely from that forbidden crossover that both science and religion can derive a practical, modern efficacy. Although scientific knowledge could be spoken of as value-free and culturally neutral, it operated in late-nineteenth-century India with an evangelical zeal that was derived from the belief that science showed the true path from backwardness and superstition to enlightenment and Christian efficiency. In turn, Christianity obtained its civilizational task from the imperatives of technological progress, so that Christian efficiency was not at all a contradiction in terms, but a powerfully naturalized ideology.[2]

At the level of their self-understanding, and much of their rhetoric, scientists and missionaries did, indeed, contradict much about each other. If we stop at noting the logical contradictions in their discourse, however, we miss seeing the intricate ways in which their material practices and, indirectly, their discourses of progress often worked to facilitate rather than to impede one another. Their lack of discursive coherence could conceal this functional symbiosis, but the full consideration of their official pronouncements as well as their everyday activities gives us a richer picture of the relation between religious and scientific modernity.

By the end of the nineteenth century, British imperialism in India was more strongly committed to being an agent of civilizational change than to governing in harmony with indigenous beliefs. The East India Company's policy of religious neutrality and its attempt to govern through a noninterfering "rule by custom" had by mid-century been replaced by the Crown's explicit desire to civilize through direct, authoritative rule by law.[3] The success of a practical, modernizing model of improvement did not occur, however, through abandoning the strongly Christian proselytizing spirit that was sanctioned by the Evangelical movement in the early nineteenth century. Although the Evangelical movement itself had not succeeded in its aim of converting India, its emotionally charged sense of historic duty invested even secular modernizing projects in the late nineteenth century. Conversely, missionary work, too, became strongly "modern."[4] Thus, what we often think of as diametrically opposed spheres of intellectual and material practice, namely, the religious and the scientific, were not in practice mutually exclusive. Neither scientific nor religious discourse can be understood solely in the sphere of its own self-conceptions, which are often articulated in abstract philosophical terms; each must also be understood through its social activity. In examining the mutual dependence of the ideologies of secular modernization and of religious salvation in the political and economic context of advanced imperialism, we explore within yet another sphere the saturation of modern world views by scientific models of progress.

The Charter Act of 1813 granted missionaries the right to establish churches and conduct religious activities in India, a right for which they had been vigorously campaigning, against the East India Company's pragmatically Orientalist policy of preserving the "ancient usages and institutions" of the country.[5] Missionaries now became central in the development of Indian educational institutions, and played a leading role in decisions related to curricula, pedagogical philosophies, and language instruction through the first half of the nineteenth century.[6]

Whereas from the 1820s to the mid–1850s, a predominantly religious agenda motivated colonial education, after Thomas Babington Macaulay's 1835 Minute on Education, and more strongly in the Crown period (after 1858), the Utilitarian approach gradually superseded the religious and Orientalist approach. Macaulay, for example, strongly influenced by Francis Bacon's philosophy, emphasized the separation of the political and cultural from the moral and religious. His Minute on Education gave official sanction to the redefinition of civilizational progress in terms of increasing the sum of private happiness, which in turn could be ensured by ensuring the security of individual citizens and private property.

The Education Dispatch of 1854 by Sir Charles Wood made explicit the connection between higher education and the creation of a middle class disciplined in the new structures of manufacture and consumption that imperial trade necessitated.[7] The dispatch recommended the establishment of different levels of education for the masses and for the upper classes, explicitly linking education, production for imperial markets, and the creation of a new class structure that piggybacked on, but significantly rigidified, existing stratifications between the lower and the middle classes.

Class stratification and production for distant markets were made possible not only by scientific secularization (understood as necessarily entailing a hostility to religious or moral agendas) but also by an ideology that wove religion and science together. The colonial relations between religion and science, and between morality and political economy, were not characterized so much by religion and morality giving way to science and economics as by a synthesis and redefinition of terms. This redefinition was crucially guided by modern scientific and technological systems, which must nevertheless be understood as incorporating or working with (rather than eliminating or replacing) nonscientific or premodern discourses.

By the mid-nineteenth century, scientific thought was rapidly becoming the privileged paradigm for both political and cultural discourse. Some might argue that it remained simply a strategic but subordinate add-on to social models of progress that missionaries held—but it would have to be shown, then, that missionaries managed to keep their models of civilizational progress free from the influence of eighteenth- and nineteenth-century scientific thought. The evidence seems to favor the possibility that, by the late nineteenth century, scientific categories fundamentally changed the ways many religious missions conceptualized their own civilizational tasks.

On the side of administrative policy, it is equally unclear that by the end of the nineteenth century, a practical and scientific imperative had superseded or displaced the moral and religious. As late as 1880, Richard Temple, a prominent colonial administrator, wrote an overview of British policy in India in which he defended the policy of state support for Christian missions on the basis of a moral and practical argument.[8] As Viswanathan has shown, early-nineteenth-century hopes that Christianity would gain a whole heathen nation of converts had given way to a more pragmatic model of social change. Temple's account, too, shows this.

> In the early part of the nineteenth century . . . high hopes prevailed in the Christian world regarding the conversion of the heathen in India. Glowing visions floated before the imaginations of pious and enthusiastic men. . . . It is now seen that the conversion of the heathen, though steadily continuous, will be comparatively slow, and that *Christian teaching* must, with *education* as its handmaid, wend onwards a persevering way, through not only the thick masses of heathenism, but also the new and many-sided beliefs which advancing knowledge and civilization may be expected to produce.[9]

Here we see an administrator, with no affiliation to any mission, offering a hierarchy in which religion precedes science: Christian teaching, supported by its handmaiden, literary and scientific education, is the moral beacon for colonial India. But by this time, toward the end of the century, secular education, as many commentators have noted, had more or less won the official battle for pedagogical priority. As a government administrator, Temple (unlike missionaries of the time) clearly had no strategic goals in terms of validating religious over secular instruction. How did Temple understand the value of Christian instruction in such a way that it did not threaten to invalidate the benefits of secular scientific pedagogy, and so that he could still see it as valuable?

By the late nineteenth century, Christian missions had incorporated the promise of scientific civilizational progress into their projects in a way that made the workings of both scientific reason and global capital integral to their self-conception. Thus by the end of the century, instead of discarding religious missions as a bygone ideological practice, administrators continued to recognize them as the moral engines of civilizational progress.

Temple describes a need for two kinds of missionaries in India: one kind to work among the "heathen wild races" and "lower orders" of society, and to run practical projects such as "public teaching, school work and administrative business"; the other kind to win over upper-caste, highly educated Hindu "opponents, more subtle and more formidable than any yet encountered [by missions]." He emphasized that the latter grade of missionary was to be relieved from administrative work and teaching, as they had to put all their skills and energy into fighting philosophical battles. "Such missionaries are those who have been despatched by the Cambridge . . . and the Oxford mission. . . . They will meet with Natives highly educated in the Western manner, into the recesses of whose thoughts some entrance must be won by a careful and gradual approach."[10] This second form of mission work was influential in the establishing of literary curricula in universities, and became progressively more literary in an academic sense, distancing itself from the mundane operations of the economy. The former, however, was explicitly practical in its outlook. In the upliftment programs of the lower castes and classes, Christian proselytizing and the arguments of secular, scientific, and economic modernity coexist.

Rather than a supersession of science over religion, we can detect a synthesis of the contradictory elements that, despite its local incoherencies, in effect empowered the resulting universalizing discourse by incorporating the potential for a strategic flexibility of style and argument.[11] The Basel Mission, active in North Canara and Malabar, offers a rich example of this phenomenon.

The Basel Mission Society and Technological Modernity: Interconnected Constructions of Work and Progress

The Basel Mission Society was begun on 26 September 1815 as an evangelical mission society. Although the mission was headquartered in Basel, Switzerland, and until the early twentieth century was mostly staffed by Swiss-German missionaries, it operated in accordance with the guidelines of the British colonial state in India. Its records thus add to our understanding of the scholarship on colonial missionaries, reminding us that the civilizing mission was not an isolated British phenomenon. Local people did not distinguish the Basel Mission in terms of its national origins from other missions; in fact, in 1912 an Indian official in Tellicherry attributed the success of the Basel Mission to "British perseverance." [12]

In August 1834, missionaries Samuel Hebich, Johan Christopher

FIGURE 11: "The missionary, in this case a German Lutheran, with some of his flock." From Michael Edwardes, *British India 1772–1947: A Survey of the Nature and Effects of Alien Rule* (London: Sidgwick and Johnson, 1967), 367.

Lehner, and Christian Lenhard Greiner arrived in Calicut and soon established headquarters at Mangalore. By 1900, they had set up missions along the southwestern coast of India, mainly in North Canara and Malabar.[13] Their most successful manufacturing units were weaving and tile-making, and they had an active printing press.

Basel missionaries were soon actively involved in diverse aspects of local life and scholarship. Hermann Gundert, a Basel missionary, compiled the first Malayalam dictionary. L. J. Frohnmeyer wrote a Malayalam grammar and several textbooks on science, and was conferred a fellowship by the University of Madras in recognition of his scholarship. The Basel Mission Society set up tile, dyeing, and weaving factories "where technical knowhow was imparted to the locals by German experts."[14] These were managed by the Basel Mission Trading Company (later incorporated into the Commonwealth Trust Limited), which also established mission presses at Balmatta and Mangalore for publishing "spiritual and educational books."[15]

They began work in the 1830s, and by 1882 the mission had set up, as the central component of its project, an industrial mission, with a mercantile branch office in Basel, Switzerland; its funds were managed by a joint stock company, which paid the net profits back to the mission as a "free contribution." The mission set up factories in which converts were trained in industrial skills so as to be economically self-sufficient. This had the benefit of producing profits that were channeled back into mission work, thus making the mission financially viable, and at the same time creating alternative forms of labor for populations that had been in the equivalent of agrestic slavery, or tribals whose livelihoods had formerly been outside mainstream economic networks but now were becoming less and less viable as a result of state controls on forest use.

The Religious Significance of Work

Basel missionary J. Müller described the importance of the mission's industrial undertakings, which were managed on sound business principles, as follows.

> [O]ur workshops are not asylums for the lame, the blind, and the crippled, and still less are they an El Dorado for sponges and idlers. They are workshops where the work done is adequately paid for, and where the wage is in most cases honestly earned. . . .
>
> The purpose is not only to offer needy converts an opportunity of earning their livelihood, but also to train them in diligence, honesty, and steadiness of character. . . . This is no small achievement in a land where the dignity of labour is unknown, where indolence and mendicancy are regarded as no disgrace, while on the other hand mechanical and manual labour are considered degrading. We have to combat ideas and weaknesses which the habits of centuries have ingrained in the character of the people.[16]

Müller is writing in the early twentieth century, by which time the Mission Society was a well-established industrial mission. In his rhetoric, the linking of Christian morality with the ethics of work makes use of a dichotomous representation that was ubiquitous in the eighteenth and nineteenth centuries: that of the upstanding, hardworking citizen of an industrial nation as opposed to the uncivilized, lazy non-Western native. The positive value of work is a central motivating element of this representation, as is the marking of different peoples in terms of their capacity for disciplined and efficient productivity. The idea of virtuous labor is the link that interweaves moral and religious with practical and scientific discourses.

The connections between work, morality, industrial production, and civilizational progress are unambiguously laid out in the Basel Mission reports. In 1873, an annual report noted: "Not only have a large number of our converts and inquirers honestly earned a comfortable livelihood, but the order and regularity to which all the labourers are trained exercise a very beneficial moral influence on them."[17] Reminding us that this method of salvation through work was consistent with the moral teachings of the Bible, an 1887 annual report noted: "The principle acted upon in these [industrial] establishments is, 'If any man will not work, neither shall he eat.' 1 Thess. 3, 10."[18] And an 1879 report gave a detailed explanation of their aims.

> The object of these industrial pursuits, as we have frequently stated in our Reports, is both directly and indirectly to further and strengthen the Mission cause. . . . We may briefly distinguish between three contiguous objects: first, a *philanthropic* one, i.e. to provide employment and honorable means of subsistence to many Christians as well as probationers, who otherwise would hardly know what to live upon. Second, a *pedagogical* object: to train our Native Christians to habits of regularity and steady, honest labour, and thereby to raise them both socially and morally. Third, a *civilising* object: to benefit the country at large by creating a class of Christian artisans, mechanics, tradesmen, etc. Fourth, a *financial* object, to find new sources of revenue.[19]

Here, working for a reward is represented as a Christian virtue; thus in training the natives in disciplined habits of work, the missionaries saw themselves as serving God. Further, the Basel Mission Society was not ignorant of the necessity of a laboring class in order for the country to produce goods and services reliably. Such a reliable, predictable labor force could only be produced if the natives were persuaded to give up their love of idleness in favor of a life of virtuous labor.

The Christian emphasis on work and labor and corresponding proscription against idleness was by no means only a nineteenth-century phenomenon; historians have documented that it can be traced back to at least the thirteenth century. Anson Rabinbach has noted that "from the thirteenth until the middle of the nineteenth century Christian writers, ministers, and middle-class moralists all accorded idleness an esteemed place as the nemesis of an orderly life and the discipline of work. . . . Thus, when the old Christian proscription against idleness was reinvigorated at the end of the eighteenth century, it was usually invoked as a criticism of the persistence of older preindustrial behaviors often directed against the new industrial order."[20]

As in the case of biblical ethnography, which experienced a revival in the eighteenth century as European travelers were faced with a bewildering variety of new peoples, a Christian moral condemnation of idleness found new resonance in the postindustrial revolution climate of industrial production. Christianity's campaign against idleness among peasants and the new working class in Europe gained an added normative dimension in the colonies, where Christian missions were faced with masses of heathen, among whose sins against the moral order idleness was only one. In the colonial Christian literature, then, we see the native constituency of the missions as the subjects of a simultaneously racialized and classed discourse. These normative discourses were instrumental in producing members of a new laboring class, who remained racially marked but were now accorded the possibility of transcending their racialized non-Christian identities by virtue of their new position in the international division of labor. Work was the way out of heathen backwardness.[21]

The Social and Political Significance of Work

The attempt to uplift rural and tribal populations into a Christian, moral, and industrious way of life had more than just religious significance. The groups that were explicitly targeted by the socially active missions were most often those that lay outside of the classes that had already been incorporated into the new economic networks. As Viswanathan has noted, the already existing stratifications in Indian society had made it possible for the incorporation of many existing upper- and middle-caste groups into various levels of the new economic systems, through their rigidification rather than structural alteration.[22]

However, low-caste landless laborers, untouchables, and tribals were not considered readily assimilable without being schooled into the systems of labor required by the new production requirements. Christian missions targeted these groups for two reasons, which they explicitly articulated. First, the more oppressed a group was by indigenous structures of power, the more willing its members were likely to be to abandon those structures for a new religion. Second, these groups appeared to practice all manner of sins against Christian values, and thus were most radically in need of guidance from above; this guidance had to take the shape of instruction in the virtues of cleanliness, industriousness, and Christian worship.

Evidence for these two claims can be found in churches' and missions' representations of their goals and achievements in the late nineteenth and early twentieth century. The report of the Third Decennial Congress of Protestant Missions, 1892, reports that the major part of Christian conversion throughout the nineteenth century came through "mass movements" among lower castes and aboriginal tribes.[23]

The bishop of Madras, reviewing the work of the church from the 1880s through the beginning of the twentieth century, wrote:

> Tyranny and oppression had prepared [outcastes] to seek consolation from Christ. Five-sixths of the converts of the last forty years come from the depressed classes. . . .
>
> Experience shows a marvelous capacity for improvement latent in the outcastes. In a single generation the children of parents living as drunkards and thieves, in filthy surroundings, eating carrion, are turned into cultivated men and women. Within thirty years, from villages once given over to awful orgies of sacrifice, the streets running blood, Christian people have proved ready to walk fifty miles to take reverent part in a service.[24]

This representation of outcastes is rendered in the standard model of savagery. The constituency of the church was thus marked by caste, and their representation always fed off racialized tropes of barbarity. They were transformed, through upliftment programs, into groups newly marked by their (working-) class status. A large and well-trained laboring class in the colonies was essential for the functioning of the empire; the social work of missions directly contributed to the creation and schooling of this work force.

Thus the work of the Basel Mission had a direct influence on caste occupations (and intercaste relations), and served an important purpose in the context of the national political economy.

Work, Religion, Politics: Connections between Christianity and Technological Modernity

I have suggested that the terms in which religious virtues, Christian morality, and civilizational progress are represented have direct consequences for the creation of new social relations and of new systems of labor. As Raymond Williams has pointed out, changes in social relations are interlinked with changes in the mode of production. "New social relations, and the new kinds of activity that are possible through them, may be imagined but cannot be achieved unless the determining limits of a particular mode of production are surpassed in practice, by actual social change."[25] Discursive formations might constrain the imaginative power of individuals, even nations; we cannot adequately explain social change, however, without paying attention to actual changes in production practices. The Basel Mission Society not only brought a Christian discourse to the tribal and lower-caste populations of Malabar; they also brought the factory system and modern work routines.

The process by which an existing mode of production is superseded by a new one is both a basic and a superstructural change; but further, the complexity of the accompanying social changes makes it clear that no assumed dichotomy between basic and superstructural, or institutional and ideological processes can be sustained. Clearly, both the rhetoric and the practices of the Basel Mission had direct ideological and material consequences not just for individual converts but also for social relations and production methods in the whole region.

Most historical scholarship on mission education has focused on the intellectual traditions it fostered. The case of the Basel Mission shows, however, that, in addition to the intellectual content of missionary ideas, there were material conditions created that brought large numbers of previously agrarian or tribal people into a new system of production. Church missions, particularly the Basel Mission, were actively involved in bringing about the social, cultural, and material conditions for the shifting of emphasis from production for subsistence to production for abstract markets. The Basel Mission's Industrial Committee, "recognising that industrial enterprise lacks a solid basis without business management," sent a businessman to Mangalore in 1852, who took over the purchase of raw materials and the sale of the finished product. Missionary J. Müller explained: "To understand the Basel Industrial Mission aright we must keep in mind the fact that we have to do not with industrial schools [which now 'belong to the past'], but with industrial establishments, that is, with the factory system."[26] The point emphasized here is that mission work does more than merely educate people who will later join a work force; rather, it produces that work force at the same time that it educates and preaches. The rationale for this was phrased in standard business terms: "experience has shown that with the keen competition of foreign countries the day of the individual workman is over, and only an industry with modern machinery and a whole sale business is capable of holding its own against competition."

Of course, the endeavor was not practical and technological to the exclusion of the moral and religious imperative. "The moral influence of work on the formation of character is sustained and deepened by daily religious instruction. Before beginning the day's work morning prayers are regularly read by the manager or a native pastor. Not only does the whole work thus receive a certain consecration, but the employees see that their employer is interested in their spiritual welfare." And again, we are reminded that practical, modern management is an ally of Christian activity: "We consider it . . . important that our Christians should learn to provide for their own church and schools, and for the sick and poor . . . , and to contribute . . . [to] the spread of the Gospel. In order to realize this aim, savings banks are organized . . . and the people are urged to make good use of them."

Most missions in rural south India pitched their proselytizing and social work at "outcastes" and tribals. In the Christian institutional writing of this period (late nineteenth and early twentieth centuries) there is constant reference to "mass movements," a term that refers to large-scale conversions to Christianity. These occurred most often during and after famines or epidemics. Such movements occurred in Madras, Travancore, Bengal, and areas of central, northeast, and northwest India, among "depressed classes" and "aboriginals."[27] Many missions began industries because new converts either had no livelihood or lost it upon conversion, or because they were forbidden by their new priests from continuing in their former "un-Christian" trades. The Basel Mission was among the most successful of all the south Indian missions in their industrial projects.

The Christian Missionary Society and Malabar Tribes

One of the best-known missions in south India was the Church Missionary Society, active mainly in the princely state of Travancore, south of Malabar. Henry Baker of the Church Missionary Society regularly picked up runaway slaves (bonded laborers). He is reported to have traveled through forests doing the "good work of evangelizing and civilizing which he carried on in the teeth of many difficulties and perils" from 1849 on.[28] Reverend George Matthan wrote in 1870 that Baker "received special funds from his friends in England to appoint a person to go to the mountains and jungles in search of such people. He got a volunteer to do that job on a small allowance just for his food and clothing."[29] Baker became familiar enough with indigenous tribes to publish a book in the scholarly ethnographic style of the time, entitled *The Hill Arians of Travancore*.[30]

Baker's work among, and apparent sympathy toward, untouchables and tribals was a source of annoyance to forest officials, who complained that the Hill Arians (Mala Arayar) destroyed valuable teak and blackwood trees. Forest officials tended to favor an aggressive disciplinary approach toward them, rather than the seemingly unjudgmental, Christian friendliness that Baker claimed to offer them. There was constant friction between the Mala Arayar and government officials involved in forest work. In 1856 the Mala Arayar mounted an attack against "government officials who used to catch people to carry cardamom from the hills to the plains with no wages."[31]

By 1879 there were over two thousand Christian Mala Arayar. The raja of Travancore, initially hostile, had become hospitable to the mission under persuasion from the British Resident.[32] There was always conflict between the mission converts and the upper castes, whose traditional power over their laborers' bodies was being eroded by the mass conversion movements. Nairs and Syrian Christians (the already existing Christian population, which claimed Hindu upper-caste roots) often tore down mission sheds that functioned as churches and schools for Pulayas and Parayas (whose main occupation had been bonded labor for the former groups), or beat up those who attended the institutions. Baker, in 1866, wrote on the Pulaya's improved condition: "Many of them, since becoming Christians, are getting lands registered in their own names, build better houses, are better clad and . . . are clearing a large extent of marsh land for paddy, all on their own account, and in *no* way helped by us, except in arbitration of boundaries between them and some Syro-Romanists and Nairs."[33]

Missionary activity did, to a certain extent, draw the government's attention to the condition of the lower castes. During the famine of 1861, Travancore durbar physician Dr. Wahring and missionary Charles Mead (of the London Missionary Society) opened a Pulaya Charity School, and they received the Travancore government's assistance for providing meals. In 1867, Travancore legislation was passed prohibiting the eviction of tenants from land. Converted laborers often refused to work on Sundays. They were commonly beaten by their employers, and occasionally Church Missionary Society missionaries took up their cause with the dewan. One missionary writes in 1860 that "many slaves escaped floggings by fleeing to the hills, . . . carrying with them the knowledge of salvation through Christ, to their fellow slaves."[34] He gives a glowing account of his missionary society's social work, noting that although the mission had only seven Europeans in Travancore, they had made great progress among the "aboriginals" of the forested regions of the Western Ghats. Once again (as in the epigraph to this chapter) we see the metaphors of wilderness tamed, and barren desert made productive.

> Moondakayam is a station formed amongst the hill Arians. A short time ago the wild elephants wandered unmolested in their native

> jungles, tigers and leopards were sights of everyday occurrence, . . . one was obliged to carry a gun . . . the jackal and the wild boar were too mean enemies to be noticed amid the savage wildness of the forests. . . . The desert has indeed blossomed as the rose. A neat church and Bungalow have been erected; a colony has been formed, and betwixt six and seven hundred natives have been received into the church by baptism, as the first fruits of the coming harvest.[35]

All this indicates that even the Church Missionary Society, whose methods were not as technologically sophisticated as the Basel Mission's, targeted oppressed sections of the population and intervened in the prevailing mode of production to the extent that those sections were imbricated in new production processes and relations.

Despite these efforts, the social effect of Church Missionary Society projects was limited by the long-term pragmatic choices they made. These choices were closely linked to their vision of social and scientific change, in which they were not as technologically utopian as the Basel Mission. Toward the end of the century there were only twenty pupils enrolled in the Pulaya Industrial school in Kottayam, largely due to pressure from Nairs and Syrian Christians against the school. Most Church Missionary Society educational institutes (the CMS College Kottayam, for example) were attended predominantly by Syrian Christians.[36] There is only one instance of a Pulayan being admitted to the CMS College; this "resulted in such a commotion that the boy was obliged to leave for the . . . German [Basel] mission,"[37] where he later became an ordained minister, an elevation in status he would not have gained in the Church Missionary Society. There were also many instances of Pulayas leaving the Church Missionary Society to join the Salvation Army, another mission that emphasized the imparting of practical skills and new trades to converts from lower castes.

Scientific knowledge helped missionaries to gain converts in other ways. The second half of the twentieth century saw many active medical missionaries, such as Ida Scudder in Madras and Vellore, and Leith in Travancore. Famines (notably those in 1837–1838, 1877, and 1901) created opportunities for relief work and preaching. At the 1879 Decennial Conference of Protestant Missions it was noted that "The medical missionary often secures entrance and acceptance where a preacher or teacher would be immediately rejected, . . . opening up the door for the evangelists to preach the gospel."[38]

It is clear that most missions used scientific and technological knowledge on a day-to-day basis, most often in strategies for winning more converts.

The Salvation Army, for instance, actively used the ethnographic science of race, and worked with the police in characterizing tribes as criminal, in accordance with legal definitions. Criminalized tribes, assigned to the care of the Salvation Army mission, would then be moved to agricultural settlements or found employment in factories.[39] With most missions, and the Basel Mission in particular, science was more than just an instrumental strategy to facilitate conversion; it was an integral part of their model of progress. In the Basel Mission reports and literature, whenever the basis of the mission's work was described, science, rationality and economic success were used as markers of true progress.

If we look at the functioning of religion, economics, and technology at social levels other than the official government representations and upper-caste responses, a complex picture emerges: one in which secular, moral, and economic agendas combine in complex and sometimes contradictory ways; in which moral and religious discourses were actively deployed in support of a capitalist mode of production, which in turn necessitated or at least made possible newly restratified social relations. For the late nineteenth century and the early twentieth century, particularly with respect to the activities of Christian missions among economically and socially underprivileged sections of the population, in rural and tribal areas, it is difficult to sustain a consistent notion of a split between secular knowledge, helpful to a capitalist division of labor, and moral and religious knowledge, morally opposed to political and economic exploitation.

Mixed Modernity: Implications for Understanding Colonial "Scientific Modernity"

Both religion and science are components of the mixed forms of modernity that were not aberrant but necessary features of the organization of colonial society. The discourses of religion and science, while contradictory at the level of official representations, have significant interests—material and ideological—in common. These interests become apparent when we examine how they jointly form the conditions of production for the practices and discourses of labor among low-caste and tribal populations.

Both the self-representation and the practices of Basel Mission Society missionaries indicate that there did exist a mode of mission activity that was not necessarily at odds with practical and technological activity. Although this is most clearly evident in the work of the Basel Mission, other socially active missions too, in targeting marginalized or exploited sections of the population, effectively created a new laboring class, as they invariably found that they had to create new livelihoods in order to facilitate conversions. These new occupations they found for their converts were structured in the context of a national and global political economy.

If such a mixture of the scientific with the moral and religious is fundamentally constitutive of colonial and postcolonial modernity, as I claim, we should expect to find it in current representations of progress.[40] Indeed, we see the strands of moral and scientistic argument preserved in recent histories of Protestant mission activity.

A historian of the Basel Mission Society, writing for the *Basel Mission Souvenir* in 1980, sums up the specificity of the mission's achievement. "Nearly 150 years ago, the Basel Missionary Society . . . initiated . . . their missionary work in Malabar Coast. . . . The Basel Mission Society not only aimed at preaching the Gospel but laid equal emphasis on *economic upliftment* of the under-privileged section of the then existing society. . . . Basel missionaries emphasized not merely the literary meaning of the word conversion, but focussed on the need to make a departure from a state of traditionally-held superstitious beliefs to *rationalistic thinking* and *reality-oriented living*."[41] It is emphasized here that the important achievement of the Basel Mission Society was its introduction of economic rationality, through which it brought the underprivileged sections of Indian society into modern ways of thinking.

Reverend Alagodi, writing in 1973 about Protestant medical missions, reveals the moral charge with which the transition from savagery to modernity is naturalized, in an appeal that simultaneously invokes the gospel and science. Medical work, he says, "helped in breaking down the prejudice of those who would not otherwise be willing to listen to the Gospel. It helped them to gain the affection of the people and to rescue the thousands of poor people from the murderous grasp of their miserable native quacks."[42]

This is Alagodi's modern commentary on this phenomenon, and replicates the divisions between Western/Christian and native/heathen on the basis of the opposition between science and superstition. Such cross-fertilization of discourses and practices continues to characterize diverse representations of modernity. They can function together despite their apparent inconsistency precisely because of the practices that undergird these discourses. In the late twentieth century, of course, they are refracted through the specific production relations and social stratifications of postcolonial India, and can by no means be read directly from colonial models. As in the case of anthropology, however, an examination of the conditions of production of knowledge in the late nineteenth and early twentieth centuries can form a basis for a critique of postcolonial knowledges and practices if it is combined with a history of discourse and practice through the twentieth century. Such an analysis would be considerably different from the extension of a colonial discourse analysis to postcolonial society, as it does not postulate an essential, unchanging style of representation but holds that representations and discourses of science, morality, and progress must be read along with their refractions through historically specific practices.

Thus the historical roots of scientific and technological modernity are multistranded; they arise not simply from supposedly protoscientific thought but also from multifarious, seemingly contradictory, political, economic, religious, and cultural discourses. Rather than see the turn of the nineteenth century as a period in which the force of scientific thought and technological planning replaced older forms of social practice (either colonial or indigenous), we might profit by paying attention to the processes by which older discourses were interlinked with or incorporated within newer scientific modes of thought and practice. A historiography that recognizes the interdependence of discourse and practice is more analytically flexible and offers more satisfying explanations of social change than one that relies solely on the analysis of ideas, however powerful those ideas might be.

Dialogues with the Past: Paying Attention to Practice

Historian E. P. Thompson has argued that the Industrial Revolution fundamentally changed the rhythms of social practice as it altered the technological and economic systems within which people had to live. Thompson has shown how Methodism provided an inner compulsion to work that reinforced the new, automated, clock-regulated work regime and successfully fed the need, in post–Industrial Revolution England, for low-paid wage labor.[43] He identifies Andrew Ure's *Philosophy of Manufactures* as the most complete articulation of what was to become the "economist case for the function of religion as a work-discipline." Ure envisioned the factory as a "vast automaton," whose main obstacle was not a technological problem but the problem of "training human beings to renounce their desultory habits of work, and to identify themselves with the unvarying regularity of the complex automaton."[44]

Anson Rabinbach, discussing Thompson's description of eighteenth-century work disciplines, comments: "The literature of labor in the first half of the nineteenth century was characterized by this traditional attitude but with a new twist: idleness was literally as well as figuratively the primary sin against industry."[45]

In nineteenth-century Europe, a number of supplementary discourses on labor and idleness emerged, connecting them to class, race, climate, and gender. These discourses traveled to non-European contexts with interesting modifications. With exploration and colonialism came the need for categories by which to describe and know native non-Western populations. Rabinbach notes this development, and places it in the early nineteenth century. "The attack on idleness was zealously applied to both Europeans and non-Europeans. Ethnologists were struck by the *horror laboris* of the 'uncivilized races,' where 'idleness and savagery' were synonymous. . . . At the beginning of the nineteenth century, ethnographers attempted to demonstrate that civilization not only promoted the habit of industry, but encouraged physical development as well."[46]

The connection between work habits and morphological characteristics provided the obvious link to generalizations based on racial groups. Anthropometry was a vigorous science in the colonies, and tribes or castes were often categorized as suitable or unsuitable for labor based on both physical and supposedly inherent psychological attributes.

Rabinbach argues that the ethnocentric and ethnographic representation of non-Western peoples as barbaric gave way in the last third of the century to "more scientific" representations of the body as labor power, divorced from its context: "a more scientific evaluation of work, often materialist in emphasis, gradually displaced the old moral discourse"; and "The long tradition of industrial edification in which the moral, intellectual, and spiritual benefits of work are opposed to the debilitating effects of sloth and the endemic laziness of barbaric peoples, is beginning to lose its discursive power [in the last third of the nineteenth century]."[47]

This periodization is accurate if one restricts its purview to intellectuals in Europe alone. In the colonies, however, through the turn of the century and well into the twentieth century assumptions continued to be made of the inherent incapacity of primitive people to grasp the forces of modernity. Such assumptions were often mixed in contradictory ways with a scientistic discourse of improvement through energetic efficiency, of the kind that Rabinbach sees as displacing the older religious moralizing model. Rabinbach's periodization risks ignoring, however, the important ways in which older thought forms continue to exist within and inform newer conceptualizations.[48]

Paying attention to European scientific discourses and labor practices,

Rabinbach assumes that a particular discourse drops out of existence when it can no longer be discerned in these accounts. This is a methodological error, because in the nineteenth century, especially in the case of discourses around labor, Europe is not an analytically complete category of historical analysis. The position of Asian and African colonies in a global system of production allowed labor practices to soften, or liberalize, in Europe even as they hardened in the colonies. The colonies occupied a structural position defined as sources of raw materials and labor for an imperial political economy. The hardening of labor practices in the colonies was illustrated in the discussion of plantation labor in chapter 4. A more complete global picture is readily available in the growing literature on colonial history, a considerable body of academic scholarship that is inexplicably missing from Rabinbach's otherwise fascinating study.[49]

At exactly the time when Rabinbach sees the representation of idleness as morally repugnant and anti-Christian drop out of labor discourses, we find evidence, as this chapter has shown, that the instituting of modern technologies of labor among tribals and low castes in south India was permeated by Christian moralizing discourses. Let us briefly reconsider the work of missionary Samuel Mateer.

Reverend Samuel Mateer of the London Missionary Society traveled extensively among the hill tribes of Travancore. His book, *Native Life in Travancore* (1883), describes the shifting cultivation of the Kanikar (also known as Mala Arayar) as the mode of cultivation that uses the least amount of labor for the desired agricultural output. The Kanikar are unmindful of the destruction of valuable forest lands, he writes, as they move on to another plot when one area is exhausted. "[T]heir migrant habits arise partly from laziness. . . . They prefer this savage life, but should be encouraged to settle if possible: only by such means can they be reclaimed to civilization and education."[50] Mateer's amateur ethnographic observations lead him to associate savagery with idleness. He repeats the argument made by ethnographers and foresters regarding the sedentarization of nomadic populations, implicitly appealing to both a Christian model of morality and the anthropologically legitimated hierarchy that placed industrial sciences and settled agriculture above nomadic pastoralism in a scale of civilizational modes.

Science, Modernity, and the Privatization of Property

The characteristics of Kanikar lifestyles are laid out in Mateer's account as evidence of their insufficiently modern systems of knowledge. Although they possess a vast store of knowledge about their surroundings (such as the whereabouts and traces of wild animals, the means of making fire from bamboo and reed, building bridges out of bamboo and rattan, and so on), this body of knowledge is deemed useless, except in guiding sportsmen to their game or warning travelers against danger.

> Though thus familiar, from ages of experience, with the ways of the forest, these poor people are not gifted with even an ordinary amount of knowledge, not one of them being able to read or write, except very recently a few in Pareychaley Mission district, who have learnt to read a little and to sing Christian lyrics. They can never tell their own ages, and if asked, sometimes make absurd guesses. They are unable to count to a hundred; over ten they lay down a pebble for each ten. They knot fibres of various climbing plants to express their wants. At Purattimalei twenty years ago, only one had seen a white man before; none had ever travelled a greater distance from home than twenty miles.[51]

The requirements of modernity were, then, literacy, mobility, and familiarity with the West or Westerners. Mateer, of course, was speaking from a context in which the print media, global trade, and cosmopolitanism were fundamental accompaniments of progress. Missionaries were, like any other social actors, products of their time and place, and in articulating their models of progress necessarily employed the practical and scientific definitions of modernity that circumscribed their own worldviews.

To point out that both tribal and missionary worldviews were specific to their context, however, is not sufficient to explain why one of them became dominant. Precisely because of their mixture of spiritual and scientific worldviews, the missionaries were able to claim that whereas the natives were prisoners of their contexts (knowing their immediate surroundings intimately, but ignorant about the wider world), the colonizers were possessed of knowledge that was context-independent and universalizable. This assumption of the context-independence of their knowledge then made it possible for them to assert their own moral and ethical superiority over all of the native "races." "The fate of the hill kings [Kanikar, Mala Arayar] is rather sad. For ages past they have boasted of being the undisputed lords of the primeval forests. The elephant and the tiger were their only foes; but with snares and traps they could hold their own against these enemies. But they could not resist the onward march of a superior race."[52]

Implicit in the rhetorical move of embedding the natives in the specificity of their surroundings was the implication that the alleged superiority of the newly arrived race was derived from its power over not only these same surroundings but also over any conceivable space it was likely to come across in its onward march. Modern scientific knowledge could transcend locally specific knowledge because of the power of generalization, by which the local became an instantiation of the global. Although local detail was sometimes empirically useful in filling out these instantiations, it would remain always weak and backward in comparison with global models that homogenized space and difference.

Modern scientific thought won out not merely because it could synthesize general laws from local detail; to suggest that would be to make the idealist error that many critics of post-Enlightenment science make when they suggest that modernity is necessarily epistemologically violent because of its reductionist appropriation of local cases into a global whole.[53] To understand the "success" of modernity we must look at the processes by which new modes of thought and new modes of production were simultaneously established. This simultaneity of ideas and production practices was no coincidence; these processes were constitutive of and dependent upon each other.

David Harvey has analyzed post-Renaissance changes in conceptions of social and geographical space, and the corresponding changes in social practice. He suggests that the conceptualization of space as homogenous is only the first step in consolidating its modern use, the crucial additional step being the introduction of private property. It is only with the second step that older forms of social practice are rendered irreversibly obsolete, impracticable, and backward.

> The conquest and control of space . . . first requires that it be conceived of as something usable, malleable, and therefore capable of domination through human action: . . . space as abstract, homogeneous, and universal in its qualities, a framework of thought and action which was stable and knowable. . . . But there were islands of practice within a sea of social activities in which all manner of other conceptions of space and place—sacred and profane, symbolic, personal, animistic—could continue to function undisturbed. It took something more to consolidate the actual use of space as universal, homogeneous . . .—the pulverization and fragmentation/privatization of land as property.[54]

The construction of homogenizing or essentializing discourses is an admittedly important, but not sufficient, step toward the establishment of a modern social order. Ideas and ways of knowing become hegemonic always in conjunction with practices that alter the way a society produces its means of sustenance, and reproduces its social order. It is these ground-level changes in practice that make older practices and epistemologies unviable.

As we saw in chapters 2, 3, 4, and 5, the imperatives of colonial production necessitated the widespread introduction of private property in areas that had formerly been organized according to social and trade relationships along regional group formations; these were by no means inherently harmonious, and were specific to factors such as caste interactions, tribal occupations, the power and distance of the ruling king, and geographical location. The introduction of private property regimes "pulverized" space, in the sense that older patterns of natural resource use became subordinated to an institutionalized set of universalizable assumptions about the management and development of nature. In that these ideas were considered universalizable, and were presented and legitimated by the standards of an objectivist context-independent model of scientific knowledge, they can be traced in both metropolitan and colonial contexts. It is the specificity of their operations in the colonial context, however, that makes these ideas relevant to the construction of modernity in colonized societies. This specificity cannot be historically analyzed if we ignore the contradictory forms of scientific and moral practice.

The analysis of modern forms of knowledge should be inseparable from the conditions within which these forms were produced and which enabled them to operate with validity. Universalizable knowledge about the environment could incorporate "facts about" local knowledge only within a system that gave priority to freely exchangeable units of property, of labor, and of information; within this system there would be no space for the now apparently nonrational cultural beliefs that had earlier framed the local knowledge systems. Thus in the case of the foresters in chapter 3, local details were very important factual components of the science of forestry; nevertheless, the context in which local knowledge had been acquired and systematized by tribal population was effectively superseded by the new conditions of production. Tribal populations were, in these new conditions, appropriated as sources of information about the use and collection of forest products, and as sources of labor in forest production systems. In the process, much of the local procedures of knowledge acquisition were rendered unviable or disrupted.

The analysis of colonial discourse offers us insights into the epistemological assumptions on which representations of colonized peoples were based. Discourse analysis does not, however, always explore the material basis of such assumptions. In other words, the ground-level functional coherence of these discursive contradictions is not, indeed cannot, be investigated if high discourse alone feeds the analysis. Conversely, the idea of contradiction as constitutive of modernity (and, by way of modernity, of postmodernity) is naturalized if we analyze high discourse divorced from practice. That is, contradiction as such is seen as an inevitable, natural feature of post-Enlightenment social orders. The exploration of the political function of these contradictions (their functional incoherence) is then closed off.

It is insufficient to speak of changes in belief systems, or discursive epistemes, without speaking also of the changes in political and economic systems that enable, and are in turn legitimated by, those new forms of thought. Faced with a vast empire populated by communities they considered barbaric, European intellectuals and administrators did not really sweep tradition out abruptly with the broom of modernity. The widely accepted model has modern political economy replacing Christian, moralizing worldviews, thus ushering in modernity. This implies a separation between modern science and Christianity. I have suggested that this was an idealist myth, often transgressed in practice. The myth was indeed a powerful one, and often influenced policy decisions and institutional practices. Rather than simply tracking the path from official discourse to the resulting practices, however, we must also move in the reverse direction, reading the practices on the ground that transgressed official ideals. Such transgressions should be regarded not just as aberrations but also as spurs to investigating the mutual constitution of scientific and nonscientific spheres of practice.

CHAPTER 7

A Global Story

IMPERIAL SCIENCE RESCUES A TREE

The great attention which English botanists of the early part of the [nineteenth] century had given to taxonomic questions and the problems of geographical botany was largely the outcome of the growth and expansion of the Colonial and Indian empire during these years. This expansion was so notable a feature of the time that it almost obsessed the minds of the scientific workers who saw the strange flora opened to them, full of deep interest and presenting problems of the greatest importance, both botanical and geographical.[1]

The history of plants in the nineteenth century—their classification, collection, transplantation, cultivation, and commodification—is interwoven with some of the central issues of colonial science studies, environmental history, agricultural economics, and the political economy of local knowledge. This chapter studies the symbolic and ideological significance of a famous plant transfer of the mid-nineteenth century: the transplantation of the cinchona tree from the Andes Mountains of Peru to the Nilgiri Hills of south India. The colonial cinchona story is familiar to historians of botany. Its sheer scale, encompassing the policies of three colonial empires—Spanish, Dutch, and British—and involving botanical activities in disparate locations on three continents—the Americas, Europe, and Asia—makes it a classic illustration of the role of botanical networks in the nineteenth-century world economy.

Toward the end of the eighteenth century, several Spanish expeditions descended with their natural history manuals and measuring equipment upon South American landscapes with a view to filling in the constitutive details of Linnaeus's classificatory system. Other European countries followed suit. Large numbers of botanical specimens were transferred across the Atlantic to the private herbaria and public gardens of Europe. During the course of the nineteenth century, several species of plants were transferred from South America to be domesticated in other parts of the world, most of them tropical colonies of the French, Dutch, and British. The frenzy of collecting and measuring did not go unnoticed by South American governments: a Brazilian government agent said suspiciously of Alexander von Humboldt, "I never saw anyone measure so carefully land that was not his."[2]

What gave geographers and botanists the authority to measure land that was not theirs, and to collect specimens from outside their own imperial possessions? Botanical knowledge was, in this process of exploration and collection, becoming a resource (both intellectual and political) that could be deployed in the rhetoric of national, imperial, or universal human progress. I will explore how scientific knowledge served as the link between national self-interest and humanitarian service by examining one small but crucial moment in the development of British economic botany in the mid-nineteenth century.

The transplantation of cinchona saplings from the Andes Mountains of Peru to the Nilgiri Hills of south India brought the local knowledge of a subsistence economy into the global arena of free enterprise. There are many ways in which this story could be told. As historians, we make narrative choices based on our theoretical and methodological commitments. The explanatory value of combining the methods of social and economic history with literary theory can be illustrated by reconstructing a narrative of the cinchona story. I suggest, in this chapter, that we might frame investigations into the epistemological and political status of science and indigenous knowledge without seeking refuge in the comfortable positions of idealist romanticism or uncritical relativism. We might draw methodological guidelines for this from environmental and social history and from literary and cultural studies, while remaining committed to finding more complete explanations of historical and scientific change than any one of these disciplines alone has offered.

The Cinchona Story

The bark of the cinchona tree, known to the Andean Quechua-speaking inhabitants of Ecuador, Peru, and Bolivia, was used by a Jesuit priest around 1640 to cure Ana de Osorio, fourth countess of Chinchon, wife of the viceroy of Peru, when she suffered an attack of malaria in Lima. Soon after, this "Jesuit's powder" or "Polvo de la Condesa" was brought to Europe, and by the end of the seventeenth century entered the London pharma-

FIGURE 12. Cinchona leaves. From Clements R. Markham, *Travels in Peru and India* (London: John Murray, 1862), frontispiece.

copeia as *Cortex peruanus*.[3] In 1742, Linnaeus named it cinchona, "with the intention of thus immortalising the great and beneficent acts of the Countess of Chinchon."[4] The world market for quinine grew rapidly, as European imperial forces relied increasingly on this antimalarial drug to enable their tropical expansion. The Andean republics dominated the market until the late nineteenth century.[5] South American exporters employed native Andeans to harvest cinchona bark from naturally occurring clumps of cinchona trees in the dense mixed forests of the Andes.

In the mid-nineteenth century, cinchona seeds and saplings were taken from Peru, nurtured in Kew Gardens, and then transplanted to plantations in the Nilgiri Hills of South India, Bengal's Mungpoo Valley, interior mountains of Sri Lanka, and parts of the Caribbean. The Dutch also established plantations of cinchona trees in Java, and the plantation economy finally undercut the South American trade. Although this project involved dozens of botanists, explorers, administrators, and gardeners, one narrative emerged to dominate much of the representation of the British cinchona transplantation. The man who seized the public eye with his heroic narrative was Clements Markham, later knighted for his contribution to science. The following account is from his 1862 book, *Travels in Peru and India,* in which he recounts his pursuit of the cinchona tree in the Peruvian Andes.

Markham's Adventures

The cinchona-growing region of Peru was known as the Caravaya forest, and bordered Bolivia's cinchona forests. By the mid 1850s, the governments of both Peru and Bolivia were aware of the global interest in quinine, and were becoming protective of their cinchona resources. Markham writes of his exploration of the Caravaya forests: "This part of the enterprise was surrounded by peculiar difficulties, arising from the jealousy of the people, habitual with the Bolivians, and recently excited in the minds of the Peruvians of Caravaya."[6] He had hoped to be undisturbed in Caravaya, and to find the Peruvians less possessive about their trees than the Bolivians: "As a considerable part of the revenue of Bolivia is derived from the bark trade, which is not the case in Peru, the Bolivians are exceedingly jealous of their monopoly."[7] In Caravaya, however, Markham ran into Don Manuel Martel, a former member of the Peruvian military, who was making a clearing in the forest to grow sugar cane. Martel related to him the story of Justus Hasskarl.

Justus Charles Hasskarl, a Dutch botanist sent by the government of Java to collect plants and seeds for their botanical gardens, had spent over a year in Peru collecting plants.[8] In the village of Sina, near the Peru-Bolivia border, under a false name, José Carlos Muller, he requested a supply of cinchona plants from the governor. The governor refused, but introduced him to Henriquez, an enterprising Bolivian. Henriquez, whom Markham describes as "a clever and intelligent but dishonest and unscrupulous man," employed an "Indian" to collect the plants. Hasskarl left with his booty, but the inhabitants of the villages bordering the cinchona forests raised an outcry and threatened to cut off Henriquez's feet if they caught him. In 1852 the first cinchona plantation was established in Java, just south of Batavia.[9]

After telling Markham the story of Hasskarl's cinchona theft, Martel vowed that if anyone else were to attempt to take cinchona plants out of the country, "he would stir up the people to seize them and cut their feet off." Markham comments, "there was evidently some allusion to myself in his bluster." When Markham arrived at Sandia, from where he had planned to begin his collecting, he found hostile municipal authorities who, he says, "took measures to prevent me from procuring a supply of chinchona plants or seeds, influenced by motives which exposed their ignorance of political economy, while it displayed their activity and patriotic zeal." Martel had, apparently, written to the inhabitants of Sandia, and was busy warning the inhabitants of villages bordering the cinchona forests to prevent the foreign explorer from removing plants and seeds. Markham reports, "My mission was becoming the talk of the whole country; and I at once saw that my only chance of success was to commence the work of collecting plants without a moment's delay, and, if possible, anticipate any measures which might be taken to thwart my designs."[10]

Although one Indian abandoned Markham's party after the first day, he was able to continue with three men, and was successful in gathering abundant supplies of cinchona plants. Deeper in the Caravaya forests, he made the acquaintance of Quechua *cascarilleros* (veterans of the bark trade, which had ended in 1847) and *collahuayas,* collectors of drugs and incense (traditional traveling healers who passed their knowledge of herbs from father to son). Markham employed Mariano Martinez, a cascarillero who had guided the French botanist Weddell on his 1846 visit, to lead him into the forest "where no European had been before."

Markham describes a long and arduous journey in search of cinchona trees, where he scrambles up "giddy precipices"; entreats and threatens (in Quechua) "mutinous Indians"; dodges wild animals, injury, and disease; and chews coca to dull his hunger. He managed to collect about five hundred cinchona plants of different varieties, which his gardener packed in "Wardian cases," specially designed to withstand the long journey to London.

On emerging from the forest, however, just as they finished packing the plants, Markham's local assistant, Gironda, received "an ominous letter" from the Alcade Municipal, obviously instigated by Martel, ordering him to arrest Markham and his guide (Martinez), and to prevent Markham from taking away any cinchona plants. Markham wrote back, quoting the provisions of the Peruvian constitution, contesting the Alcade's power to order his arrest, "concluding with an expression of my sense of his patriotic zeal, and of regret that it should be accompanied by such misguided and lamentable ignorance of the true interests of his country. Nevertheless, I felt the imperative necessity of immediate flight, especially as I obtained information from an Indian that Martel's son and his party . . . were coming down the valley to seize me, and destroy my collection of chinchona-plants." [11] Judging discretion to be the better part of valor, then, Markham escaped with his precious booty of saplings, thwarting his pursuers who, the narrative suggests, had they been successful, would have destroyed not only him and his collection but the future of the cinchona tree.

Markham's Contexts

In order to contextualize Markham's text, let us briefly look at two British colonial botany texts that illustrate, first, the significance of global exploration for the practice and theorization of the discipline of botany, and, second, some culturally specific claims associated with the pursuit of observer-independent objectivity.

In 1855, Joseph D. Hooker and Thomas Thomson published the first volume of *Flora Indica,* a projected (but never completed) systematic guide to the plants of British India.[12] They explained the pressing need for comprehensive and standardized reference works for use in medical and economic botany, as these fields in India were "at a standstill for want of an accurate scientific guide to the flora of that country."[13] Hooker and Thomson perceived the field of natural history as going through a crisis of authority. The rising number of amateur botanists and the resulting proliferation in the number of varieties reported as discovered was proving too great for the Linnaean canon, by which twelve words were allowed for a specific character.

In this introductory volume, explaining the urgent need for systematization, the authors complain about the proliferation of erroneously classified species and genera, and suggest that the lack of conceptual and consolidating work threatens the progress of the science and its commercial applications. They declare themselves impatient with "backyard botanists" who fail to understand the full range of the science, and whose view is skewed, unlike "those who extend their investigations over the whole surface of the globe."[14]

Exploration of the globe's flora had led to an explosion in number of species becoming known to scientists. This was not merely a quantitative shift but a development that prompted something of a conceptual crisis in the field of natural history. The global reach of botanical science now demanded a science that could both integrate the bewildering diversity and bring analytical rigor to the examination of single plants. Attempting to professionalize and systematize the field so as to exclude the bungling amateurs who were threatening to suffocate its system in meaningless detail, Hooker and Thomson assert, "We are . . . anxious to refute the too common opinion. . . . [T]hat descriptive botany may be undertaken by anyone who has acquired a tolerable familiarity with the use of terms. . . . to develop the rules of classification, to refer new and obscure forms to their proper places in the system, to define natural groups and even species on philosophical grounds, and to express their relations by characters of real value and with a proper degree of precision, demands a knowledge of morphology, anatomy, and often of physiology."[15]

The epistemological consequence of the new professionalism is illustrated by the comments of I. H. Burkhill, reporter on economic products to the Government of India and economic botanist to the Botanical Survey between 1901 and 1912.[16] Remarking on the move from natural history to botany, Burkhill asserts that the displacement of man from the content of knowledge (so that the scientific agent is able to have precise knowledge of the object-world of nature) accounts for the universality of botanical knowledge.

> [A] system emerged out of the Technology of Healing in a particular part of Europe, was accepted and clad there with a vocabulary of precision, as Sciences must be, and . . . this system has been spread over the World, India included, without meeting a rival. Its origin was, as it were, by a sublimation in which Man was displaced from the focus of thought that "the plant" might be placed there. . . . A writer in Calcutta claims that those who wrote in Sanskrit possessed the natural science, on the ground that they applied Sanskrit names to plants. The argument is false. Apply the test, namely which is at the centre, man or the plant; when applied it will be found that man maintains his post at the centre.[17]

Burkhill finds botany more objective than indigenous systems of plant science, as excluding the observer from the system of knowledge eliminates subjective biases and cultural meanings.

From these two treatises we see what was at stake in early botanical science—in short, systematization, professionalization, and observer-independent objectivity. Let us look briefly at its institutional context in the same period. In the second half of the nineteenth century, Kew Gardens developed from a private garden into one of the central instruments of botany and empire. Sir Joseph Banks, the amateur botanist who managed Kew before it became a state institution, was also a founder of the African Association (the precursor of the Royal Geographical Society), and a president of the Royal Society. These scientific institutions provided the infrastructure that made it possible for English explorers to map and survey the entire globe. Explorers, adventurers, and men who believed themselves invested with a spiritual or historical mission were among the first administrators, planters, foresters, and settlers of colonial India.

The Royal Geographical Society began as a gentlemen's club called the Raleigh Club, formed to bring together "the most eminent Travellers in London": "[T]he object of the Club was that travellers may assemble in social converse, who have visited distant countries, particularly those that have been little explored."[18] In 1830, along with members of the African Association, they formed the Royal Geographical Society. Its charter stated that its "sole object shall be the promotion and diffusion of that most important and entertaining branch of knowledge—geography; . . . the interest excited by this department of science is universally felt, [and] its advantages are of the first importance to mankind in general, and paramount to the welfare of a maritime nation like Great Britain, with its numerous and extensive foreign possessions; . . . its decided utility in conferring just and distinct notions of the physical and political relations of our globe must be obvious to every one."[19] Among its primary tasks was "To prepare brief instructions for such as are setting out on their travels, pointing out the parts most desirable to be visited, the best and most practical means of proceeding thither, the researches most essential to make, phenomena to be observed, the subjects of natural history most desirable to be procured, and to obtain all such information as may tend to the extension of our geographical knowledge."[20]

From an amateur group of travelers who met for conversation, a strategically important imperial instrument grew within less than fifty years. Initially not funded by the state, its members were the aristocratic elite, with wealth and leisure plus curiosity and a love of adventure. Markham was thus participating in a long history in which amateur science and the imperial state were linked through curiosity, adventure, and glory. The imperial ruling class was drawn from the ranks of exactly such men. For example, among the six members of the founding committee of the Royal Geographic Society was Mountstuart Elphinstone, who had spent most of his life as an Indian civil servant, and who had been governor of Bombay from 1819 to 1826. In 1832, an affiliated geographical society was formed at Bombay, supported by the Navy surveyors.

The fourth president of the Royal Geographic Society, William Hamilton, characterized geography as a selfless search for universally beneficial knowledge.[21]

> The real geographer becomes at once an ardent traveller, indifferent whether he plunges into the burning heats of tropical deserts, plains, or swamps, launches his boat on the unknown stream, or endures the hardships of an Arctic climate, . . . Buoyed up in his greatest difficulties by the consciousness that he is labouring for the good of his fellow creatures, he feels delight in the reflection that he is upon ground untrodden by man, that every step he makes will serve to

> enlarge the sphere of human knowledge, and that he is laying up for himself a store of gratitude and fame.[22]

Through amateur scientific practices and organizations, an institutionalized science emerged that was explicitly imperial, yet simultaneously represented itself as a universal, humanitarian project. Exploring such simultaneously nationalist and universalist claims in the context of Markham's description of the cinchona transplantation, I argue that, rather than seeing them simply as pseudo scientific rhetoric, we can use these apparently contradictory moments to understand the political and economic interests undergirding the accumulation of scientific knowledge.

Analyzing Environmental Narratives

The diverse aspects of the cinchona story have been told before.[23] I revisit it here not to tell a new story but to explore one particular strategy of analysis: that of reading environmental histories as literary, cultural, ideological, and political narratives.

Precisely because history is multiply narrativized, the environmental stories we tell remain dissatisfyingly incomplete when we read off just one master explanatory narrative from our sources—whether our favorite choice for dominant explanand is the economy, politics, intellectual trends, or the personal psychology of the men behind the events. Reading environmental history as a multiply constituted discourse allows us to ask questions such as what rhetorical strategies were employed here, to what effect, and through invoking what imaginary audience? Which ideologies are made explicit, and which ones smuggled in, and why are those choices made? How are changing tensions between environmentalist protection and capitalist development negotiated? What equivalences or differences between nature, history, progress, and peoples are postulated? By invoking specific ideological formations within the questions about rhetorical strategies we ensure that although we are reading environmental theory and practice as discourse, we are not sliding down the slippery slope to pure textualism.

Narratives, of course, are stories—but no story is only "just-so"; stories have a cognitive value, they help us to make sense of the world and to organize our knowledge of it and our actions within it. By throwing together particular kinds of observations within particular narrative structures, stories make correlations and imply causal relationships between otherwise disparate phenomena. In rereading a system of narratives, we can unravel the threads of these implied causations, and ask how they function to structure the reader's conclusions.[24] Thus we do not have to abandon attempts to trace causal connections among, say, systems of political patronage, professed values of humanitarianism, and the advancement of botanical knowledge; nor do we have to abandon the search for explanatory adequacy.

Environmental discourse analysis profits greatly from the insights of science studies into the social construction of scientific knowledge. Through a science studies lens, the cinchona story offers insights into the professionalization of botany and geography; the role of political patronage and military expansion in the constitution of scientific institutions such as Kew Gardens and the Royal Geographic Society; and the importance of scientific practice in the colonies to theoretical advances in the natural and human sciences, as well as to the trajectories of individual scientific careers. The interweaving of scientific knowledge, political interests, and military power in the cinchona story elucidates one of science studies' central insights, namely, that science is a form of culture; scientific epistemologies and cultural politics are mutually constitutive. But further, it refines the "science as culture" insight by reminding us that the politics of culture must be studied through a close attention to the histories of institutions, networks of power, and the economics of global capitalism—categories that, despite their obvious import, are often neglected in overly culturalist analyses of science, that is, those that posit culture as the cause of scientific beliefs, without investigating the genesis of ideological formations which constitute that culture.

Overlapping with science studies, and even more important to the cinchona story, is the growing field of colonial environmental history. As Richard Grove has shown, early conservation thinking was in opposition to the interests of capital. From his history of environmental thinking in the colonies, we know that it is not enough to understand cinchona transplantation merely as a commercial, profit-making enterprise; it must also be seen as part of a century-long process in which, from 1760 through 1857, there emerged a set of scientists and doctors with explicitly articulated concerns for the well-being of nature, and with correspondingly reformist politics. The relation between these scientific experts and the British colonial state developed dramatically in the context of Indian agrarian and forestry concerns.

Within the context of early environmentalist thinking, cinchona transplantation was an important symbolic success. Historians have pointed out that British collectors chose to cultivate the "wrong" species, and were never able to compete commercially with the Dutch, who planted *Cinchona ledgeriana* extensively, successfully cornering the world market with its higher quinine yields. But the British cinchona project bore fruit that was in many ways a more complex success, in terms of linking the institutional and epistemological bases of colonial power. As a grand scheme for the more efficient nurturing of crop plants through the application of modern scientific methods, in the service of the state and of humanity, it created symbolic capital that was successfully invested in the creation of networks of scientists, institutions, and administrators that linked British imperial colonies through botanical gardens and scientific departments in the metropole and peripheries. As Richard Drayton points out, "The ultimate long-term result of the cinchona scheme would be the building around Kew in the 1880s and 1890s of networks of colonial gardens which would culminate in the great Imperial Departments of Agriculture for the West Indies and West Africa."[25]

Kew became a center of imperial botany from which were made decisions far exceeding the importance of botanical classification. Directors of Kew from Joseph Hooker onward were to make decisions about policy and personnel in far-flung regions of the empire, not only in the field of economic botany but also all through the overlapping domains of colonial politics and metropolitan entrepreneurship. Thus, although the British error in the choice of cinchona species has been commonly seen as evidence of the failure of the British to win the quinine race against the Dutch Empire, the significance of the cinchona adventure can be fully appreciated only if we read all its multiple ideological meanings.

To agrarian and political economists, the cinchona story is an element in the establishment of colonial plantation economies in Asia and the Caribbean. Nevertheless, it has been postulated as an apparent anomaly in the common plantation narrative of private capital versus local sustainable agriculture, according to which monocultures displaced local modes of land use, and caused the immiseration of indigenous labor. Lucile Brockway, in one of the first comprehensive accounts of the politics of quinine, suggests that the environmentalist notions that rhetorically undergirded the cinchona project resulted in labor relations and land-use policies that were less devastating to local communities and soils than the unchecked capitalism of the other south Indian plantation strategies.[26] The Nilgiri Hills of south India, a primary area of cinchona transplantation, occupied a romanticized ecological space in the colonial imagination, unlike the economic and managerial conceptions of the plains landscapes or the fearful, Christian antipathy for the chaos of jungle areas.[27] South Indian hills were the site of many privately managed, commercially successful plantation crops: tea, coffee, and rubber, for example. Cinchona plantation, on the other hand, remained the domain of government rather than private investment, and was framed in terms of humanitarian benefits, public health interests, sustainable land-use policies, and the incorporation of hill tribes' agrarian systems into the colonial economy. The "hill station" became a popular site for wildlife preservation, the protection of traditional agriculture and tribal forest rights, anthropological primitivism, public health concerns, and the creation of English landscapes.[28] Thus in order to understand why the political economy of cinchona transplantation to the Nilgiris was different from that of tea or rubber plantations, we must read the multiple, contradictory discourses of environmentalism, primitivism, and romanticism.

The narratives of nature, then, might be successfully read through the lenses of multidisciplinary analysis, combining literary studies, environmental history, and science studies. If we borrow the methods of close textual readings and discourse analysis from literary studies, and pay attention to long-term ecological effects and the politics of environmentalism, while elucidating the mutual constitution of scientific knowledge and the material conditions of culture, we will have a methodological framework that allows us insight into the multiple significations of representations of nature.[29]

Reading the Cinchona Narratives

Travels in Peru and India is a book of over five hundred pages, filled with not only detailed discussion of the logistics and economic viability of the cinchona project but also reflections on the responsibility of colonizers toward the colonized and of the scientist toward nature.

Although this text is from the mid-nineteenth century, science and political economy were closely linked long before this time. This was especially important for a world power like England. In Elizabethan England, geographical knowledge had been systematized for the benefit of both explorers and scholars. Seventeenth-century explorer Richard Hakluyt worked for the East India Company, and used his geographical knowledge to draw up lists of commodities that could be bought or sold at various ports around the world. Clements Markham saw himself in the tradition of "adventurers" such as Raleigh and Drake, who were, he says, "fathers of our science" in a time "when it was the highest ambition of the flower of England's sons to add to her fame by achieving discovery in distant lands."[30] He was secretary of the Royal Geographic Society and wrote the official history of it for its fiftieth anniversary in 1881.

When, by the middle of the nineteenth century, it was becoming evident that there was a danger that a South American monopoly over quinine might be established, it seemed only logical to call for an explorer to seek a solution. The problem was complex, involving several different fields of knowledge and activity. It called for an explorer with an understanding of the role of adventure and discovery in the maintenance of empire; but it also called for someone who was both botanically and linguistically well versed. The explorer needed not only to be able to identify the particular species and variety of plant that would provide a commercially viable yield of quinine but also to be able to converse with the local inhabitants in order to elicit information about the whereabouts and accessibility of the plant specimens. Aside from exploration, the project included the envisioning, planning, and execution of a massive program of transcontinental transplantation. The key resources, then, were located on three continents: South America had the plant, but Britain did not have South America; Britain, however, had India, which had, the geographers estimated, a climate that would be hospitable to the plant; and in London itself, there was the pivotal resource, Kew. Kew Gardens would serve as incubator to the plants uprooted from their native South American soil, nursing them to a state sufficiently vigorous enough to withstand a second voyage and a subsequent transplanting in foreign soil. Kew was a midway resting point for the plants on their voyage from Peru to the Nilgiris; moreover, it had the professional gardeners, the greenhouses, and the expertise that India lacked. Kew also marked cinchona's point of transition from uncultivated rain-forest growth to commercial cash crop.

In addition to its obvious economic import, such a transformation carried with it considerable symbolic significance in the context of a scientific discourse of progress that dominated much of nineteenth-century colonial thought. Progress could be measured by the advancement along a trajectory defined by several oppositions. Societies could be arranged on a scale from savagery to civilization in terms of their advancement from social, political, and biological chaos and wildness toward cultivated orderliness; from locally effective power to global penetration; from passive effeminateness to active, powerful virility; from concrete and empirical forms of knowledge to the abstract and theoretical. Thus, although the empirical knowledge of the healing properties of *quinaquina* bark may have been with the native Andeans for hundreds of years, it remained trapped in the forests of the Andes, and was, in this form, "useless" to the rest of the world.

Markham does not count indigenous knowledge of cinchona as scientific, even though, almost every step of the way, he is guided, tutored, and sheltered by indigenes. Markham asserts that even though there is evidence

FIGURE 13. The Palm House, Kew Gardens. Photo: K. Philip.

that the "Indians" had long known about the medicinal properties of cinchona, "they attached little importance to them." There seems little evidence for this assertion, however, as Charles Marie de la Condamine, Joseph de Jussieu, and Hippolito Ruiz had all reported, in the eighteenth century, that the Indians had taught the Spanish the use of the bark. Alexander von Humboldt and Richard Spruce had both described elaborate indigenous systems of healing (based on the "hot" and "cold" properties of various illnesses and cures), in which cinchona bark had a specific place.

Markham records that "the first description of the cinchona tree is due to that memorable French expedition to South America, to which all branches of science owe so much." This was made by La Condamine, "the first man of science who examined and described this important plant." Thus a colonial history of quinine could only, by definition, recognize the first description of the cinchona tree to have been made by a European scientist, even though native South American bark collectors had been carrying the bark around in their medicinal pouches for some time. How are we to understand the distinction between scientific and nonscientific description here? The sentences in La Condamine's description did not, in themselves, constitute scientific knowledge; thus, the same words spoken by a Peruvian native would not have counted as a scientific description.[31] These same sentences in the context of the explicitly scientific purpose of the expedition, its institutional support, and its framing in terms of already established paradigms of research constituted something that was recognizable as scientific knowledge.

How did the La Condamine expedition meet these criteria? Hecht and Cockburn point out that preceding Amazon explorations had been explicitly religiously or politically motivated, whereas this one, sponsored by the French Académie des Sciences, aimed to resolve questions about Newton's theories regarding the shape and size of the earth. It included ten "natural philosophers," including La Condamine. "La Condamine's journey differed from earlier ones in that it was sponsored by a scientific institution and in principle concerned the accumulation of pure knowledge; but his botanical descriptions of plants had very practical consequences, and changed the region forever. Rubber, quinine, curare, ipecac, and copaiba oil made their entrance into European history, first as exotica and minor trade novelties, and later as the basis for substantial economic enterprises.[32]

Although the institutional context for the production of truth here is provided by scientific institutions and commercial production, it is no less important to note the rhetorical basis of truth. What rhetorical strategies were employed to establish the boundaries between scientific truth and unscientific ignorance? Let us note Markham's metaphors describing humans' interactions with nature.

Reading Markham's memoirs in the context of the models of colonial natural resource management that were developing during the nineteenth century in the official administrative literature, we can discern an interesting aspect of the metaphors he uses. Markham is cast in the role of intrepid explorer pitting himself against the elements in order to wrest the cinchona tree from the depths of the Peruvian wilderness, but this struggle is portrayed more often in lyrical than in agonistic terms. More important, nature was something that had not so much to be battled against as battled *for*.[33] Markham the modern scientist was represented as the defender of nature against the ignorance and greed of less advanced societies. For example, Markham quotes the Spanish botanist Ruiz's protests against the practices of the bark collectors of Loxa. "[He] declared that it was very injurious to the trees, many having been destroyed by it . . . thoughtlessly destructive. . . . They often pull up the roots, while the annual burning of the slopes, and the continual cropping of the young shoots by cattle, assist the work of destruction. It is therefore well that the *C. Chahuarguera* and *C. Uritusinga*, the earliest known and among the most valuable of the chinchona trees, should have been saved from extinction by timely introduction into India."[34]

Markham repeatedly draws our attention to the contrast between the respectful and conservative attitude of European science toward nature and native brutality toward nature. The image of the ignorant, destructive native (aided by his beasts) is in deliberate contrast to the nurturing hand of European science. Further, this language suggests, scientific knowledge is employed in the service of all humanity, while native interests are by definition narrow and short-sighted. Markham complains, "The collection of bark in the South American forests was conducted from the first with reckless extravagance; no attempt worthy the name has ever been made either with a view to the conservancy or the cultivation of the chinchona trees; and both the complete abandonment of the forests to the mercy of every speculator, as in Peru, Ecuador, and New Granada, and the barbarous meddling of Bolivia, have led to equally destructive results."[35] Markham portrays nature as under attack: its assaulters are those ignorant of science, while its defenders are those who can exploit and sustain it at the same time. It is possible here to disentangle two kinds of ignorant, unscientific behavior—one that is ignorant of methods of conservation and regeneration of valuable natural forest growth, and another that is ignorant of the logic of the free market. The two are connected in interesting ways.

The "barbarous meddling" in Markham's complaint refers to Bolivia's attempts to restrict the removal of cinchona bark by nonlicensed collectors and foreign nationals. A conflict over national sovereignty, ownership of resources, and access to international markets is here construed as a conflict over how to treat nature properly—Bolivia's legislation is said to have wreaked havoc on the cinchona tree. The shortsightedness of both individuals and governments is explained as stemming from their lack of awareness of the fragility and value of the natural wealth they are tampering with, or their lack of scientific knowledge of nature. If Bolivia had pursued or allowed more advanced nations to take over the systematic and scientific development of its resources, the argument goes, the cinchona plant would not be threatened, as it is now. The present embattled state of the cinchona tree, however, calls for drastic measures. In order to support his claim, Markham cites other explorers such as Ruiz, who had characterized the bark-collectors of Peru as "thoughtlessly destructive." As Peru and Bolivia lacked the capability to nurture the tree as was needed, it was a good scientist's duty to rescue the tree from the danger of extinction. What ensues is an action-packed chase drama through the forests of Peru and Bolivia, with Nature, or the cinchona tree, as the damsel in distress, to whose rescue the gallant knights of European science rush, pursued by the misguided Bolivian nationalists who seek, from selfish motives of national profit, to thwart this mission.

Markham's writing is a fascinating and sometimes inexplicably self-contradictory mixture of botanical evangelism and unabashedly commercial profit-and-loss talk. Although the book is clearly framed in the rhetoric of a conservationist agenda, in the details of the story are embedded explicit declarations of the economic motives of this project. For example, early in his narrative Markham points out that the real fear of European governments is not for the ultimate survival of the cinchona tree but for the regular supply of the alkaloid for the use of their troops and administrative staff stationed in tropical climates. "The danger . . . is not in the actual annihilation of the chinchona trees in South America, but lest, with increasing demand, there should be long intervals of time during which the supply would cease, owing to the forests being exhausted, and requiring periods of rest."[36] While he is explaining the scientific interest in cinchona, he notes that the question of introducing the plant into other countries "and thus escaping from the entire dependence on the South American forests has long occupied the attention of scientific men of Europe." The science and the economics of cinchona cultivation were thus obviously intertwined from the very beginning.

Richard Spruce, one of the collectors in the British cinchona expedition, an eminent botanist who had spent seventeen years of his life in the Amazon, put the real issue clearly when he exclaimed, "How often have I regretted that England did not possess the Amazon valley instead of India!"[37] Markham was aware, of course, of the potential commercial and military importance of domesticating the cinchona tree, and that the French and the Dutch were both attempting to cultivate it.[38] But the political reality of the vested imperial interests in cinchona cultivation did not stop Markham from representing the transplantation project in terms of universal human benefit. He tells us that the efforts of colonial powers to redistribute resources to their economic advantage is only a fulfilling of the logic of civilization. "The distribution of valuable products of the vegetable kingdom amongst the nations of the earth—their introduction from countries where they are indigenous into distant lands, with suitable soils and climates—is one of the greatest benefits that civilisation has conferred upon mankind."[39]

Moreover, the tree, if wisely conserved, is a resource that lasts forever; thus the transplantation of a tree, and the institutionalization of the corresponding systems of conservation and management in the colonies, serve as an eternal monument to the benevolence of the colonial state. Markham argues that the redistribution of natural resources, in addition to "ensur[ing] material increases of comfort and profit," has effects that "are more durable that the proudest monuments of engineering skill." He notes: "It is by thus adding to the sources of Indian wealth that England will best discharge the immense responsibility she has incurred by the conquest of India, so far as the material interests of that vast empire are concerned. Thus too will she leave behind her by far the most durable monument of the benefits conferred by her rule."[40]

Although Markham's own text is generous to the historian in terms of the internal contradictions that allow us to read the discourse of universal progress along with the discourse of economic rationality, other documents of the time make even more explicit connections between cinchona and imperial rule. In 1879, the superintendent of the Government Central Museum, Madras, Surgeon-Major Bidie, wrote to the assistant director of Kew, Thistleton Dyer, about cinchona. "To England, with her numerous and extensive Colonial possessions, it is simply priceless; and it is not too much to say, that if portions of her tropical empire are upheld by the bayonet, the arm that wields the weapon would be nerveless but for Cinchona bark and its active principles."[41]

In Markham's discussion of the applications of cinchona, he notes that in addition to "the commercial point of view," there are "motives of humanity" that ought to move coffee planters to plant cinchona for the treatment of their estate employees. Finally, he adds that he also hopes the natives will start cultivating cinchona in their gardens, as they do coffee. In order of importance, then, he has: "commerce," or national profit; private enterprise (for Europeans in the tropics); plantation labor (which, if healthier, will work more efficiently, and show higher productivity); and, as a possible incidental spin-off, the adoption of home-grown quinine by natives. The last never happened. The first two did, to a certain extent, although British quinine was never as successful as Dutch on the international market. The cinchona transplantation project succeeded in making quinine available for military and administrative personnel in all the tropical British colonies, but was a failure both as a profit-making cash crop and as a public health measure in India. Yet it was spectacularly successful as a symbol of the benevolence of both science and empire, and as a result of this served to legitimate the influence of the imperial botanical expert in colonial agrarian activity, and to make Kew Gardens the powerful center of a whole network of colonial botanical gardens by the late nineteenth century.

In the 1878 *Report of the Cinchona Committee*, Colonel Campbell Walker addresses various criticisms of the cinchona project that have been brought up in Parliament, by private investors, and in public discourse. To the demand that the government process Nilgiri bark locally so as to make a cheap febrifuge (the alkaloid source from which quinine was extracted) available to the native population, Walker responds that "a procedure of this nature would be of no benefit to the natives of India for whom a cheap febrifuge is so much desiderated," since they would be denied access to the cheap product anyway, by "speculators" who would "buy up the febrifuge and send it home [i.e., to England] to manufacturers for the extraction of quinine," selling it there at a profit. Thus, he argues, the product would automatically find its true market price, and the natives would be no better off under this scheme, while the state would be worse off, the speculators having made the profits. He states, "We do not believe in the ability of any Government to cheapen artificially the production of a commodity of public utility. But in demonstrating that cinchona can be profitably cultivated, the Madras Government has exercised an enormous influence in stimulating production, which in due course will have a decided effect in regulating the supply of bark and reducing the price of alkaloids, not for India only, but for the whole world."[42]

Such an argument assumes that market forces, like scientific knowledge, will "naturally" benefit all humanity. The easy slide from "native" to "India" to "world" is significant in this quote. The power of both the new botanical networks and the commercial market model derive from their globality. In such a global narrative, the local disparities, some preexisting, some produced by the functioning of these global processes, can be glossed over or regarded as cancelled out in the universal equation of the advancement of knowledge and profit. If the natives are still dying from malaria, it is not because science and economics have not done their best; it is only a matter of time before the results of global processes will penetrate every locality. That the resources for the production of quinine (indigenous peoples' knowledge and land) had come originally from some of these very local contexts could not count since, in such a generalized scheme, all contexts are equivalent, and must follow the laws of the market.

Thus the rhetorics of economic and scientific morality are hard to separate in the narratives both of cinchona's domestication (in India) and of its discovery and rescue (in South America). In the South American context, the cascarillas are described as having access to the mechanical aspects of cinchona bark harvesting but not to the science, and this is supposed to explain their nonconservationist habits.[43] The examples we are given of local use of the tree date from the period of the intensive South American bark trade, beginning in 1820, when European demand for the bark had risen enormously and private collectors, cascarilleros, had entered the forests in large numbers. The depletion of the bark that ensued was halted in 1834 through legislative intervention by the Bolivian government, who tried to establish a regulated system whereby all bark would be picked for the state, which would supervise the international trade. This early attempt at statist resource conservation was met with outrage by European governments, who protested what Markham refers to as a "most barbarous legislation" based on "a system of protection and monopoly." Yet, in Britain's colonies, early environmentalism was unabashedly statist.[44]

Protectionist economic practices, then, and nonconservationist, unscientific attitudes were portrayed as alike in their barbarous consequences. In this case, of course, the consequence of both was the same: less cinchona bark, at a higher price, for European nations.

Economic progress was defined as the establishment of a global free market for natural commodities; the means by which the British, French, and Dutch acquired Peruvian and Bolivian cinchona seed, however, was not exactly within the provisions of such a market system. What constituted the legitimation for smuggling the cinchona tree from Peru and Bolivia? The fact that each of the expeditions that collected specimens of the tree traveled under the aegis of science, which was a pursuit of knowledge that knew no national boundaries. Once the natural object had been transformed into a commodity, however, it became subject to the laws of the market, and here national boundaries could be legitimately invoked. Thus the rhetorics of universal scientific benefit and economic self-interest operate side by side in Markham's narrative, their separate narratives preserving a logical incoherence.

Noting the successful transplantation of cinchona in the Nilgiris and looking forward to "similar happy results . . . [in] other hill districts of Southern India," Markham suggests, "Thus will the successful cultivation of the quinine-yielding chinchona-plants confer a great and lasting benefit upon the people of India, as well as upon the commerce of the whole world."[45] Here the imperial scientist-explorer is the agent of civilization; through him nature's benefits are conferred upon mankind. The scientist speaks and acts on behalf of nature, and is willing to go to great lengths to defend it against the actions of those less scientifically savvy than himself. Thus we are to understand that the local skirmishes over property rights that mark the initial phase of the cinchona project are fought—although they might contravene the laws of less advanced nations—in the name of all humanity, including the natives of colonized nations, who, although not truly able to know and to nurture nature, are still worthy of receiving the fruits of civilization. Therefore the natives of India are repeatedly invoked as the potential beneficiaries of the cinchona project. They will enjoy the fruits of scientific labor even while remaining, themselves, ignorant of science. Colonialism, thus, is the agent of science and the spokesman for nature.

Where do the native Andeans fit in this scheme? They were not invoked as beneficiaries of the transplantation. They were not, after all, British colonial subjects; they were, however, the first link in the information-transfer chain. Interestingly, Markham never denies or obscures this position. Neither Markham nor any of the European explorers before him had been in any doubt about the fact that native Peruvians who lived in and around the cinchona tracts knew the medicinal properties of cinchona bark. Furthermore, whenever Markham sought information or guidance, it was from a native. Although colonial scientists often did not deny that natives could see natural entities, they suggested that they saw them not as continuous elements of an overall order but rather as entities invested with cultural (subjective and mythical) meanings. This knowledge, while often empirically accurate in some ways, would thus not be considered error-free, as it was irreducibly social.

One of the main differences between indigenous knowledge and a science framed by the language of nineteenth-century natural history was the apparent position of the observer or knower with respect to the object of knowledge, namely, nature or the natural world.[46] The human observer, according to the ideal of natural history, ought to be invisible or irrelevant: any educated scientific reader, in any part of the globe, would by definition know a plant once provided with the precise description of its features and its position in the taxonomic grid. Local knowledge, on the other hand, is by definition impossible without the experience of living within the natural system concerned. Thus for an indigenous Peruvian to know, say, the woodlands in France, would be inconceivable within this system. This distinction between local and global knowledge about nature helps us avoid the assumption that native or indigenous people are somehow inherently eco-friendly, as it reminds us that knowledge (local or global) is inextricably bound up with specific practices and global networks. Further, to understand that botanical transfers of the nineteenth century originated and developed in the service of particular economic and political projects reminds us that the apparent oberver-independence of universalist scientific knowledge is an illusion that is sustained through discursive contradictions such as Markham's triumphalist plunder narrative or the Cinchona Committee's free market humanitarian narrative. These discursive contradictions must thus be seen as functionally incoherent. Further, rather than see this incoherence as evidence of any essential undecidability undergirding all narrative, we can now see it as motivated by global networks of commodification that could be successfully crafted only in conjunction with such an environmentalist discourse.

Cross-contextual knowledge became possible only in conjunction with the transnational flow of men and money; along with colonialism emerged an international exchange economy in which knowledge and capital could move rapidly together across the globe in the service of increasingly efficient production. The power of the scientist lay in his supposed ability to transcend the specificity of place and time and to fashion a knowledge that posited itself as universal. The force of such a claim depends in large part on its being taken as transhistorically true. The scientist then appears both omnipotent and dispassionate, being detached from all constraints of politics and history. The scientist's supposed transcendence of the constraints of place and history are in direct contrast with the representation of the native as rooted in a specific geographic and social location. However, as I have suggested, the representation of scientific knowledge as transcending place and time was a move that always served to obscure the very specific rootedness of this knowledge in a particular political economy and set of cultural practices.

By demonstrating that behind every claim of scientific universality there lie sociopolitical enabling conditions that are rendered invisible by a language of precision and context-independence, we can begin to understand the structures that underpin this asymmetry between indigenous and colonial, or local and global knowledge. Both native and imperial knowledge derive their force from the specifics of place (or geography) and time (or history) through which their level of efficacy emerges. To assert this is not to claim that native and imperial knowledges were equivalent, or equally powerful. Imperial knowledge clearly had greater instrumental efficacy: it straddled several contexts, and crossed (dramatically, in this case) national boundaries; it did things in the world and it had historical effects that indigenous knowledge never did. But this was precisely because imperial scientific knowledge was undergirded by objective structures of power: economic and political networks that spanned continents, so that it was possible for enormous material resources and skills to be deployed by European nations in order to extract information and natural resources from another continent and transport it back, and transform it into a further resource for the building of imperial political and economic power. In other words, the claim to universality was not merely a rhetorical move made by imperial knowledge. Systems of knowledge cannot be understood solely as rhetorical constructions, but must be understood within the economic and political system that forms the frame for their conditions of production.

One way to recast this claim is to say that we can write environmental history as a history of commodities. However, I am not arguing here for a base-superstructure model in which the economy supposedly, in the last instance, explains the intellectual history of environmental thought. A history of commodities is necessarily global, as indeed environmental history must be; moreover, as I have suggested through this reading of the cinchona story, we can understand the functioning of the commodity only through a simultaneous reading of both its discursive and its material constitution.

I stress the mutual constitution of environmentalist narratives and political economy here in order to argue that this is, in fact, a rich way to understand any aspect of environmental history. Whether our historical interest is in the realm of events or of individuals, of science or of local knowledge, reading environmental history as a set of imbricated discourses allows us to understand the ways in which cultural, political, economic, and scientific narratives and practices construct and enable each other.

This approach to environmental history allows us to engage in a dialogue with the past that adds a sometimes surprising, often instructive, perspective to present-day discourses of the environment. A century is often time enough for empires to wither, and for exotic species to be hailed as ecological bedrock. In 1993, an Indian newspaper article called for local preservationists to rally around the threatened cinchona forests of the Nilgiris, urgently announcing: "Despite all the noise on conserving the biodiversity of the Western Ghats, several hundred acres of cinchona land have been cleared for planting tea, and the Tamil Nadu Government is the worst offender." Contesting the government's decision to discard the no-longer profitable forest commodity, the author employs classic environmental discourses, and chronicles the ecological splendors of the Nilgiri cinchona forests. He documents the multiplicity of species they host: Sambar deer, leopard, bear, the Nilgiri langur *Presbytis johnii,* "one of the most endangered primates in India," the floral diversity, including epiphytic orchids and the carnivorous plant *Drocera deltanta*, the avifauna, and so on. In his concluding paragraph, he recalls the words usually attributed to Chief Seattle: "We are part of the earth and the earth is part of us," and quotes extensively from this speech, including the assertion that "the white man treats his mother the earth and his brother the sky like merchandise. His hunger will eat the earth bare and leave only a desert." He exclaims: "Prophetic words indeed! The Nilgiri tribals felt exactly the same way when these hills were opened up to outsiders. Now even the remaining flora and fauna must be crying out similarly."[47]

Here the post-independence Indian state is awarded the role of the invading "white man," while the earth, flora, fauna, and tribals are cast in the role of victimized nature. Modern-day environmentalists are the audience for this call to speak on behalf of nature. The structure of this discourse, then, closely parallels Markham's; we could quite easily deconstruct contemporary Indian discourses of environmentalism for their romanticist or primitivist assumptions. As I argued in the case of the nineteenth century, so I would argue for the twenty-first: we must ask who speaks for nature, why, and what political economic networks they are caught up in. Once again, the history of commodities identifies one of the most complex environmental debates today, that of intellectual property rights, which undergirded my earlier discussion of the role of indigenous Andean knowledges.

The natural resources of tropical countries were a crucial source of the raw material that European imperial powers of the eighteenth and nineteenth centuries turned into commodities. The Dutch obtained gold and diamonds from South Africa, cocoa from west Africa, coffee from Java, sugar from the Caribbean, tobacco and rubber from Malaya; the British obtained tea, coffee, rubber, and opium from India and Sri Lanka, sugar from Trinidad and Guyana; France obtained rice from Vietnam and sugar from Haiti.[48] Most resource transfers were plantation crops; many were not indigenous, but were cash crops introduced explicitly for the benefit of imperial coffers. As Jack Kloppenburg has noted, "nearly every crop of significant economic importance—and, indeed, agriculture itself—originated in what is now called the Third World"; yet "The germplasm resources of the Third World have historically been considered a free good—the 'common heritage of mankind' . . . Plant varieties incorporating genetic material originally obtained from the Third World now appear there not as free goods but as *commodities*."[49] Identifying the "Vavilov centers of genetic diversity" (homes to most of the world's economic crops, in which he includes quinine) as lying almost entirely in the Third World, Kloppenburg argues that there has been a large-scale transfer of plant genetic resources to the First World as a result of at least four centuries of the botanical networks of conquest and trade.

This cycle of commodification of local knowledge lies at the heart of contemporary debates over biodiversity conservation, community versus corporate intellectual property rights, and patents on medicinal plants. A critically engaged history of nature can raise questions about our ongoing constructions of botanical networks, and the political economy of our own discourses about nature.

Chapter 8

Conclusion

The nature which comes to be in human history—the genesis of human society—is man's real nature; hence nature as it comes to be through industry, even though in an estranged form, is true anthropological nature.[1]

Why and how were local systems of knowing and using nature superseded by "modern" epistemological and sociopolitical systems? When we trace the operation of technoscientific modernity through its discursive and practical manifestations, we see how scientific and nonscientific spheres ideologically constitute each other. We have seen how the practices of forestry and planting physically reordered landscapes, and also how they produced ways of knowing nature that effectively entrenched a hierarchical natural order of human societies, and a corresponding hierarchy in the methods of knowing, using, and managing nature. These ways of knowing nature and natives were significantly indebted to a system of thought, anthropology, that constructed natives as objects of scientific knowledge.

The social history of forestry, the economic history of plantations, and the history of disciplinary anthropology are vigorous academic fields in themselves. In this book I have been concerned less with the debates internal to these fields than with understanding the role of science—as a resource for thinking and acting—in British colonial discourses and practices of modernity. Doing this entails questioning some commonly held assumptions about the constitution of science and modernity, and about the links between the two.

Science, I have argued, can be understood not simply in terms of "social constructions." Though valuable, the insight that science is socially constructed must be complemented with an examination of the historical specificities of the categories through which we understand the social. The cultural production of meanings and identities, the political economy and material bases of society, and the discourses and practices of nature and culture must be examined together if we are to arrive at a model that can adequately explain the totality of the role of science in society. Only a methodology that combines the best techniques of social and economic history, literary and cultural studies, and science and technology studies, while being aware of the limits of each, can be flexible and rigorous enough to facilitate such an investigation. I have tried to demonstrate the value of such a methodology by carrying out a critical study of colonial science, suggesting that a model of scientific modernity accounts for the functionality of the contradictions and incoherencies in colonial social and scientific formations.

Colonial anthropology temporally distanced natives from their observers by placing "primitive" societies in Europe's past, and then investigated them for clues about the origins of human differences.[2] These investigations commonly employed representations of natives that saw "primitives" as closer to nature, and more distant from civilization, than European societies. The metaphors that described natives as creatures of nature were more than mere rhetorical embellishments; they expressed ideological commitments, and through the epistemological frameworks they legitimated, had consequences for the practices into which tribal populations were incorporated. Ethnographic ways of knowing legitimated the management and control of tribal populations in ways parallel to the management of natural resources, with the result that natives could be construed as natural resources. Native labor was harnessed and controlled much as the tropical jungle and its products were managed and turned to profitable use through the imposition of order and predictability. The necessary connections among private property, labor, and science, within a global capitalist system of production, emerged through sciences such as forestry and anthropology, and concurrent commercial enterprises such as plantation agriculture.

The discourses around nature and natives that we visited in chapters 2 through 7 all had in common a double-paired dichotomous model of nature and culture, whose account would run as follows: the natives, being of a low or primitive culture, had failed to make the transition that modernity required, whereby cultured humans (as self-acting, autonomous subjects) separated themselves from nature in order to act upon it. High cultures separated themselves from nature so they could fully use it to the benefit of mankind, and, once having tamed, marginalized, or contained nature, they could gaze upon it with cultured appreciation.[3] Because low cultures could not work upon, transform, or efficiently manage nature, they remained on the same level as nature, thus able neither to use it productively nor to appreciate its beauties.[4]

This model powerfully structured representations of natives and nature; but obviously, its effects were not confined to the realm of ideas, as is evidenced by the imbrication of discourse and practice in the control of forest and plantation labor, indigenous use of cinchona bark, and the policing of criminal tribes. Although modern scientific practices and ideologies undergirded the dramatic changes in the lives of south Indian tribals, it would be erroneous to characterize the period simply as one of conflict between tradition and modernity, or between indigenous people and colonial science. Systems of knowledge and practice that are often represented as premodern, or as antithetical to science, were in fact not overcome, replaced, or superseded by a scientific modernity. On the contrary, earlier forms of knowing and doing entered into a complex, mutually constitutive relation with newer, more "modern" forms. Thus, for example, religious thinking and missionary practice in the late nineteenth and early twentieth centuries were not simply vestiges of an earlier, prescientific worldview; rather, religiously inflected moralizing rhetoric, proselytizing activity, the scientific justification of civilizational hierarchies, and technological production sustained each other in a matrix of interdependence.

If we are to understand the origins and legacies of the specific modes of modernity that British colonialism brought to India in particular, and to its Asian, African, and Caribbean colonies in general, we must pay attention to the contradictions and the functional incoherencies in the discourses and practices that constitute colonial scientific modernity as a totality. In the preceding chapters (especially chapters 3 and 5), I have argued for a mixed model of modernity over the model that sees modes of thought and production as succeeding, or giving way to, one another in a linear progression.

Critical Studies of Colonial Science

> For British advocates of empire, the claim of the imperial power to superior scientific understanding was seldom of first importance. It was morality, not science, that was held to justify empires.[5]

> [In nineteenth century British India], language and literature acquire an importance exceeding that of even science and technology.[6]

> [T]hough scientific and technological proficiency was often invoked by those who argued for European racial superiority, in some ways it was more fundamental than racial categories in European thinking in [the nineteenth century].[7]

Each of these three quotations represents an effort to set up a heuristic framework by which to justify the careful study of a particular historical phenomenon in relative isolation from other comparable phenomena. The first two argue for the relative unimportance of science (in comparison with moral and literary discourses) to the civilizational goals of nineteenth-century colonialism in British India, whereas the third privileges scientific over racialized discourses in the understanding of the same processes.[8] I have argued, in the preceding chapters, for conjoining rather than separating the historical analysis of these various aspects of social and cultural change.

While embracing the notion that scientific change must be studied as a social and political process, I have attempted, through an analysis of ideological and political categories, to go beyond the simple assertion that science is constructed by social categories. By problematizing the construction of those categories themselves, I have tried to demonstrate what it would mean to go a step further than noting that science is socially constructed. This step is more, however, than an infinite regress hypothesis, by which one would suggest that academics are doomed to the continual tracing of ever-expanding, equally significant, circles of social construction.

To be critical of the ways in which our society and our knowledges have been produced does not necessarily entail a radically anarchistic skepticism—that is, it is not to reject the possibility of knowing anything. Critical thought does not imply the need for a critical position outside the functioning of power—to assert that would be implicitly to employ a positivist framework in which nonsocial or pure knowledge would be available to the critic. It is from within the enabling and constraining framework of power-laden formations that we make our critical interventions, without attempting to erase our own positions within those formations.[9] But this "situated knower" must, nevertheless, be critical enough to differentiate between the myriad modes in which ideology functions, and work toward the kinds of social conditions and practices that are likely to facilitate the construction of less oppressive knowledges.

We can work toward better ways of knowing if we enlarge our definitions of science and society so that their imbrication is seen to have epistemological and political consequences. More than a decade of feminist and postcolonial scholarship has argued for a political critique of oppressive forms of knowledge and practice, a scholarly yet historically situated and politically committed critique that enables both critics and their readers to envision more egalitarian systems of knowing and doing. Sandra Harding, for example, has argued that critiques of Eurocentric science "demonstrate that equity issues are not only moral and political, as usually perceived, but scientific and epistemological as well."[10] Harding, although critical of positivist models of science and universalist definitions of rationality, does not discard the terms "objectivity" and "rationality," or cease to distinguish between better and worse explanations of nature. I concur with her in holding that we must stake out a position of strong objectivity that radically undercuts older, positivist notions of objectivity as value-free and context-independent by "thinking from" the historical particularities of different "races, classes, cultures, and sexualities." A "strong" objectivity sees these specificities not as an obstacle to true knowledge but as a "resource for those who think that our understandings and explanations are improved by what we could call an intellectual participatory democracy."[11]

A better society needs to be simultaneously worked toward (through altering social discourses and practices) and argued for (for example, on the basis of realist ontological claims against the alleged inherent savagery of natives). The separation of social practices from knowledge claims will eventually lead us to an impasse; critical studies of colonial science will do well to consider, from the start, how these two are mutually constitutive.

As Joseph Rouse has suggested, in order to develop methods of critically studying, and intervening in, both science and society, we need to be clear about the social consequences of our epistemological commitments.[12] In order to do this, not only will we have to do close, micro-level studies of scientific communities, we will also need to expand our focus to include the simultaneous processes of change in social and scientific categories. This is why (returning to the quotations I began with) I argue that we need to expand our definitions of science to such a degree that we run the risk of sometimes not recognizing where our old heuristic boundaries lay—between science, morality, economics, and politics.

The contradictions among the statements of the three quotations demonstrate that the lines between different kinds of social practice (scientific, religious, economic, or racial) are arbitrary. However, it is insufficient to append a recognition of the arbitrariness of boundaries to a business-as-usual investigation into one or another of these categories. Each of the works cited in the epigraphs offers us a good history of one aspect of science or society. However, each also misses important aspects of historical and scientific change by excluding the interconnections from its purview, thus ending up with incomplete or incorrect models of social change.

We cannot treat scientific knowledge production as a process that must be studied differently from other social practices, because of the degree to which it is imbricated with other social processes. And we can no more remain indifferent to which epistemic system wins than we can remain indifferent about what kind of society we live in.[13] Science studies, we often hear, has come of age as the newest star in the cultural studies pantheon. But before we celebrate the birth of new disciplinary constellations, we still have much work to do in dissolving the lines—disciplinary, geopolitical, and practical—that we have unquestioningly reproduced in the theory and praxis of modernity.

Notes

Chapter 1 *Introduction*

1. See William Tabb, "Globalization Is *an* Issue, the Power of Capital Is *the* Issue," *Monthly Review* 49.2 (1997): 23–24.
2. Paul Hirst and Grahame Thompson, *Globalization in Question: The International Economy and the Possibilities of Governance* (Cambridge: Polity, 1996), 9–10, cited in Tabb, "Globalization Is *an* Issue," 24. In noting this similarity between the global trends of the nineteenth and early twentieth centuries, I do not mean to suggest that, in fact, nothing has changed, nor that we can simply read twentieth-century politics directly from nineteenth-century precedents. Colonial global networks differ in many ways from postcolonial globalization, though the latter might continue to be shaped by an imperialist political agenda. For more detailed historical discussions of globalization, see Fredric Jameson, "Notes on Globalization as a Philosophical Issue," in *The Cultures of Globalization*, edited by Fredric Jameson and Masao Miyoshi (Durham: Duke University Press, 1998), 54–80; Terry Harpold and Kavita Philip, "Of Bugs and Rats: Cyber-Cleanliness, Cyber-Squalor, and the Fantasy-Spaces of Informational Globalization," *Postmodern Culture* 11.1 (September 2000).
3. Some examples are the Chilka antifisheries protests (see "Ban Soon on Prawn Culture in Chilka Lake," *The Hindu* (New Delhi), 27 June 1999), the Kaiga antinuclear agitation (see Mohammed Ahmedullah, "Sugarcoating Nuclear Power," *Bulletin of Atomic Scientists* 55.5 (September-October 1999); and the Bhopal gas-victim advocacy movement (see Upendra Baxi and Amita Danda, *Violent Victims and Lethal Litigation: The Bhopal Case* [Bombay: N. M. Tripathi, 1990]).
4. See, for example, Ashis Nandy, ed., *Science, Hegemony and Violence: A Requiem for Modernity* (Delhi: Oxford University Press, 1988); K.S.S.P., "Science as Social Activism," *Proceedings of Conference on People's Science Movements in India* (Trivandrum: Kerala Shastra Sahitya Parishad, 1984). Aspects of this vigorous debate have appeared since the 1980s in the journals *Economic and Political Weekly* and the *Lokayan Bulletin*.

5. See, e.g., Carole Crumley, ed., *Historical Ecology: Cultural Knoweldge and Changing Landscapes* (Santa Fe: School of American Research Press, 1994); and the work of biologists Richard Levins and Richard Lewontin, such as Levins and Lewontin, *The Dialectical Biologist* (Cambridge, Mass.: Harvard University Press, 1985); and Richard Lewontin, "Is Nature Probable or Capricious?" *BioScience* 16.1 (January 1966): 25–27.
6. S. Ravi Rajan, "Scientific Forestry and the British Empire," D. Phil. thesis, Oxford University, 1994.
7. Mahesh Rangarajan, "Forest Policy in the Central Provinces, 1860–1914," D. Phil. thesis, Oxford University, 1992.
8. Richard Harry Drayton, "Imperial Science and a Scientific Empire: Kew Gardens and the Uses of Nature, 1772–1903 (England)," Ph.D. dissertation, Yale University, 1993.
9. Richard Grove, *Green Imperialism: Colonial Expansion, Tropical Island Edens and the Orgins of Environmentalism, 1600–1860* (New York and Cambridge: Cambridge University Press, 1995).
10. Zaheer Baber, *The Science of Empire: Scientific Knowledge, Civilization, and Colonial Rule in India* (Albany: State University of New York Press, 1995); Deepak Kumar, ed., *Science and the Raj, 1857–1905* (New Delhi: Oxford University Press, 1995).
11. Richard H. Grove, Vinita Damodaran, and Satpal Sangan, eds., *Nature and the Orient: The Environmental History of South and Southeast Asia* (New Delhi: Oxford University Press, 1998); Madhav Gadgil and Ramachandra Guha, *This Fissured Land: An Ecological History of India* (New Delhi: Oxford University Press, 1992); Sumit Guha, *Environment and Ethnicity in India, 1200–1991* (Cambridge: Cambridge University Press, 1999); Christophe Bonneuil and Marie-Noell Bourguet, eds., "Special Issue on Colonial Environmental History," *Revue Française d'Histoire d'Outre-Mer* 86.322–323 (1999).
12. John MacKenzie, ed. *The Empire of Nature: Hunting, Conservation, and British Imperialism* (New York: St. Martin's, 1988); Gyan Prakash, "Science 'Gone Native' in Colonial India," *Representations* 40 (1992): 153–178.
13. Nandy, ed. *Science, Hegemony and Violence*; Shiv Viswanathan, *Organizing for Science: The Making of an Industrial Research Laboratory* (Delhi: Oxford University Press, 1985); Vandana Shiva, *Staying Alive: Women, Ecology and Survival* (London: Zed Books, 1988).
14. Since it is ideological constructions of "nature" and "natives" that I will address, I will drop the scare quotes from these terms, as also from "modernity" and "tribal." The terms "indigenous peoples" and "indigenes" have found favor among some writers who wish to get away from the ideological baggage of the terms "tribal" and "native." I use the colonial terms "tribal" and "native" because, first, they are consistent with the usage of the period I study; second, they are commonly used by colonial historians, without the scare quotes, and are understood to carry sets of historically laden meanings; and third, terms such as "indigeneity" carry their own particular late-twentieth-century baggage, and would not help me get away from ideologically charged terms. I will retain the

terms whose constructions I reveal even while using them, rather than seek to escape one set of ideologies by recourse to a different set.

15. For one example of the hybrid delights of science, imperial administration, and adventure, see John Keay, *The Great Arc: The Dramatic Tale of How India Was Mapped and Everest Was Named* (New York: HarperCollins, 2000). For the importance of the scientific expert to the project of planning the postcolonial state, see Partha Chatterjee, *The Nation and Its Fragments: Colonial and Postcolonial Histories* (Princeton: Princeton University Press, 1993).
16. For an example of the argument that science is essentially violent, see Nandy, ed., *Science, Hegemony and Violence*.
17. Eric Wolf, *Europe and the People without History* (Berkeley and Los Angeles: University of California Press, 1982), 17–19.
18. Such a claim is made by literary theorist Satya Mohanty, who draws on the work of philosopher of science Richard Boyd; Satya Mohanty, *Literary Theory and the Claims of History: Postmodernism, Objectivity, Multicultural Politics* (Ithaca: Cornell University Press, 1997), 144, 18–24.
19. Dominick LaCapra, *Rethinking Intellectual History: Texts, Contexts, Language* (Ithaca: Cornell University Press, 1983), 337–338.
20. Ibid., 338.
21. Edward W. Said, *Orientalism* (New York: Vintage Books, 1979); Ranajit Guha, ed., *Subaltern Studies: Writings on South Asian History and Society* (Delhi and New York: Oxford University Press, 1982–present); Chandra Talpade Mohanty, Ann Russo, and Lourdes Torres, eds., *Third World Women and the Politics of Feminism* (Bloomington: Indiana University Press, 1991); Samir Amin, *Eurocentrism*, translated by R. Moore (New York: Monthly Review Press, 1989).
22. Edgar Thurston, "Ethnographic Notes," *Madras Government Museum Bulletin* 2.1 (1897).
23. Ibid., 60.
24. Some poststructuralist narratives merely reverse this construction when they see hybridity as inherently liberating, or as worthy of uncritical celebration. They fail, like the romanticists, to look for the underlying conditions of possibility of particular modes of being and representing.
25. Edgar Thurston, "Ethnographic Notes," *Madras Government Museum Bulletin* 2.3 (1899): 133.
26. An analogous version of this romanticist construction of "primitives" is seen in certain forms of ecofeminism, which employ essentialist constructions of women as inherently closer to nature than men. See Maria Mies and Vandana Shiva, *Ecofeminism* (London: Zed Books, 1993).
27. Guha, *Environment and Ethnicity in India*.
28. Reginald G. Burton, *Sport and Wildlife in the Deccan* (London: Seeley, Service, 1928).
29. John M. MacKenzie, "Chivalry, Social Darwinism and Ritualised Killing: The Hunting Ethos in Central Africa up to 1914," in *Conservation in Africa: People, Policies, and Practices*, edited by D. Anderson and R. Grove (Cambridge and New York: Cambridge University Press, 1987), 41–61.

30. Ibid., 50.
31. Ibid., 57.
32. Ibid., 52.
33. E. P. Stebbing, *Jungle By-ways in India: Leaves from the Notebook of a Sportsman and a Naturalist* (London: J. Lane, 1911).
34. This fear lay beneath the surface of many colonial administrative concerns, as can be seen especially in official anxieties about control in the aftermath of the 1857 rebellion. See Eric Stokes, *The English Utilitarians and India* (Oxford: Clarendon Press, 1959).
35. Game conservation in the colonies has received attention from colonial historians. See, for example, David Anderson and Richard Grove, *Conservation in Africa: People, Policies, and Practices* (Cambridge and New York: Cambridge University Press, 1987); and Gadgil and Guha, *This Fissured Land*. On climate and conservation, see Grove, *Green Imperialism*.
36. Richard P. Tucker, "Resident Peoples and Wildlife Reserves in India: The Prehistory of a Strategy," in *Resident Peoples and National Parks: Social Dilemmas and Strategies in International Conservation*, edited by Patrick C. West and Steven R. Brechin (Tucson: University of Arizona Press, 1991), 40–52.
37. Grove, *Green Imperialism*, 380–387.
38. The British Crown officially took over the rule of the Indian subcontinent after the "Mutiny" of 1857. The period from 1858 to World War I is often referred to as the period of "high imperialism." Attitudes toward native colonial subjects changed considerably from the second half of the nineteenth century onward, and my discussion of attitudes toward natives should not be taken as positing a transhistorical European attitude toward the non-Western world. Rather, the perception of rising threats of native unrest, disorder, or disease in the second half of the nineteenth century marked a hardening of administrative attitudes toward local custom. See Stokes, *English Utilitarians and India*; Bernard Cohn, *An Anthropologist among the Historians and Other Essays* (Delhi and New York: Oxford University Press, 1987). Some scholars date this change of attitude to the cholera epidemic and the public health concerns of the 1830s; for a discussion of this, I am indebted to Vijay Prashad's lecture, "Colonialism and Dalit Historiography," South Asia Program Colloquium Series, Cornell University, Fall 1994.
39. See Meena Radhakrishna, "The Criminal Tribes Act in Madras Presidency: Implications for Itinerant Trading Communities," *Indian Economic and Social History Review* 26.3 (1989): 269–295; Sanjay Nigam, "Disciplining and Policing the 'Criminals by Birth,' Part I: The Making of a Colonial Stereotype—the Criminal Tribes and Castes of North India," *Indian Economic and Social History Review* 27.2 (1990): 131–164; Nigam, "Disciplining and Policing the 'Criminals by Birth,' Part II: The Development of a Disciplinary System, 1871–1900," *Indian Economic and Social History Review* 27.3 (1990): 257–287. See also Meena Radhakrishna, *Dishonoured by History* (New Delhi: Orient Longman, 2001).
40. This research was conducted at the United Theological College Archives, Bangalore, which houses the largest missionary archive in South Asia, and in the

records of the United Planters Association of South India (UPASI), the largest private plantation organization in south India. The UPASI is still very active, and has its headquarters in Coonoor, in the Nilgiri Hills.

41. For an analysis of the plantation economy in a global context, see Valentine Daniel, Henry Bernstein, and Tom Brass, "Special Issue on Plantations, Proletarians and Peasants in Colonial Asia," *Journal of Peasant Studies* 19.3–4 (1992).
42. I use the categories "scientific" and "nonscientific" in reference to the self-conception of these disciplines—that is, nineteenth-century writers in the first three categories explicitly referred to their work as scientific, whereas missionaries clearly saw themselves as non-, if not antiscientific. Planters saw scientific crop research as useful; however, they saw themselves as entrepreneurs rather than as scientists. Our task as science studies critics is to complicate these distinctions.
43. See, for instance, M. N. Pearson, ed., *Spices in the Indian Ocean World* (Brookfield, Vt.: Variorum, 1996); Philip D. Curtin, *Cross-Cultural Trade in World History* (Cambridge and New York: Cambridge University Press, 1984).
44. See Pearson, *Spices in the Indian Ocean World*, xv–xxxvii.
45. Ibid., xxiii.
46. Ibid., xxxii.
47. Hermann Kulke and Dietmar Rothermund, *A History of India* (London: Routledge, 1990), 229.
48. Ibid., 249.
49. Penelope Carson, "Golden Casket or Pebbles and Trash? J. S. Mill and the Anglicist/Orientalist Controversy," in *J. S. Mill's Encounter with India*, edited by Martin Moir, Douglas Peers and Lynn Zastoupil (Toronto: University of Toronto Press, 1999), 163.
50. Eric Stokes, *The Peasant and the Raj: Society and Peasant Rebellion in Colonial India* (Cambridge: Cambridge University Press, 1978), 36.
51. See, for example, the work of Bipan Chandra and Tapan Raychauduri, as well as the work of the Subaltern Studies collective: for example, Bipan Chandra, *Nationalism and Colonialism in Modern India* (New Delhi: Orient Longman, 1979); Dharma Kumar, ed., *The Cambridge Economic History of India* (New Delhi: Orient Longman, 1984); Ranajit Guha and Gayatri Chakravorty Spivak, eds., *Selected Subaltern Studies* (New York: Oxford University Press, 1988).
52. Gyan Prakash, *Another Reason: Science and the Imagination of Modern India* (Princeton: Princeton University Press, 1999).

Chapter 2 ***A Local Story***

1. Lord Lytton, letter to Lady Lytton September 1877, quoted in Sir Frederick Price, *Ootacamund: A History Compiled for the Government of Madras* (Madras: Government Press, 1908), 4–5.
2. Ibid.
3. Macaulay, quoted ibid. Best known as "The Whig Historian," or author of *History of England from the Accession of James the Second* (Boston: Houghton Mifflin, 1901), Macaulay was also a member of the Supreme Council of India

from 1834 to 1838, and author of the 1835 "Minute on Education" for India's Committee on Public Instruction. See *Speeches by Lord Macaulay with His Minute on Indian Education*, Selected and with an Introduction and Notes by G. M. Young (London: Oxford University Press, 1935).

4. Ootacamund is not unique in this respect; many hill stations (Simla, Mussoorie, Darjeeling, Mahableshwar, for example) were valued for their Englishness.
5. Reading resistance through the fissures in colonial texts is a well-known historiographical exercise. This is an important part of retelling hegemonic narratives of nature, although the excavation of these stories is an exercise fraught with historiographical difficulties. See James C. Scott, *Weapons of the Weak: Everyday Forms of Peasant Resistance* (New Haven: Yale University Press, 1985); Gayatri Spivak, "Can the Subaltern Speak?" in *Marxism and the Interpretation of Culture*, edited by Lawrence Grossberg and Cary Nelson (Urbana: University of Illinois Press, 1988); Jean Comaroff and John Comaroff, *Of Revelation and Revolution: Christianity, Colonialism, and Consciousness in South Africa* (Chicago: University of Chicago Press, 1991); Ranajit Guha, ed., *Subaltern Studies: Writings on South Asian History and Society*, vol. 1(Delhi and New York: Oxford University Press, 1982); Rosalind O'Hanlon, "Recovering the Subject: Subaltern Studies and Histories of Resistance in Colonial South Asia," *Modern Asian Studies* 22.1 (1988): 189–224; Jane Haggis, Stephanie Jarrett, Dave Taylor, and Peter Mayer, "By the Teeth: A Critical Examination of James Scott's 'The Moral Economy of the Peasant,'" *World Development* 14.12 (1986): 1435–1455.
6. From "The Mission of Todramala," Portuguese manuscript by Finicio, 1602, translated by W.H.R. Rivers. This is the record of a Portuguese priest's two-day encounter with the Todas. See W.H.R. Rivers, *The Todas* (London: Macmillan, 1906). Rivers hypothesizes the Todas to be "representatives of one or more of the castes of Malabar whose institutions have in some ways degenerated during a long period of isolation," or, alternatively, "one of the hill tribes of the Western Ghats who have developed a higher culture than the rest in the very favourable environment provided by the Nilgiri plateau" (*The Todas*, 716–717). The Jesuit priest's 1603 report on his visit to Ootacamund is cited in Price, *Ootacamund*, 1–2. Price notes a discrepancy in the recorded name of the Jesuit priest, who is called Ferreiri in an early translation and Finicio or Fininicio in W.H.R. Rivers's later translation (Price, *Ootacamund*, 1).
7. The 1819 "discovery" of Ootacamund is noted in Price, *Ootacamund*, 6. According to the census reports of 1891 and 1901, although the term Chetty (or Chetti) is a caste name, it is also a title, and referred loosely to petty traders and itinerant money lenders. See Edgar Thurston assisted by K. Rangachari, *Castes and Tribes of Southern India* (1909; reprint New Delhi: Asian Educational Services, 1987), vol. 2: 91–92. Poligar (or Palayakkaran) were regarded by Thomas Munro in the 1780s as "almost savage," spear-carrying feudal chiefs; by the late nineteenth century they were zamindars (land owners) who occupied "more or less wild" tracts in southern India. Ibid., vol. 6: 205.
8. Panter-Downes, *Ooty Preserved*, 30–31.

9. The importance of reproducing, through gardening, a "Home" environment can be found in much colonial literature by women. For example, Sarah Jeannette Duncan, in her 1893 novel, *Simple Adventures of a Memsahib* (Ottawa: Tecumseh Press, 1986) has her protagonist, Helen, nurture a flower garden. Helen Browne's flower garden is suggestive of all the delicacy and refinement of English life, all the purity and freshness that an English wife recreates for her family in a far-flung outpost of the empire. She "could always go down and talk of home to her friends in the flower beds, who were so steadfastly gay, and tell them . . . how brave and true it was of them to come so far from England. . . . And in the evening the smoke of the hubble-bubble was lost in the fragrance of the garden." In contrast to this gentle private garden, the India outside was desolate. It was "empty as if it had just been made"; stony, its vegetation dry or twisted, its only inhabitants "straggling . . . bands" of "brown people, . . . appearing from nowhere and disappearing in vague and crooked directions"; and beset by crows and vultures—altogether a primitive and barren landscape. Outside the bungalow was "India as it was before ever the Sahibs came to rule over it" (Duncan, *Simple Adventures*, 42, 43, 166).
10. Anthony D. King, *Colonial Urban Development* (London: Routledge, 1976), 165–167.
11. Ibid., 166–167.
12. Charles Hilton Brown 1936, quoted in Paul Hockings, *The Blue Mountains* (New Delhi: Oxford University Press, 1989). Brown was an officer of the Indian Civil Service from 1913 to 1934, and served in several administrative positions in south India.
13. Friedrich Metz, *The Tribes Inhabiting the Neilgherry Hills; Their Social Customs and Religious Rites: From the Rough Notes of a German Missionary, Edited by a Friend*, 2nd edition (Mangalore: Basel Mission Press, 1864).
14. Ibid.
15. Elizabeth A. Povinelli, *Labor's Lot: The Power, History and Culture of Aboriginal Action* (Chicago: University of Chicago Press, 1993), 11. Australia only amended the *terra nullius* clause in its constitution in the late twentieth century.
16. For a good introduction, see David Ludden, *Peasant History in South India* (Princeton: Princeton University Press, 1985).
17. Civilian, *The Civilian's South India: Some Places and People in Madras* (London: John Lane, Bodley Head, 1921), 152–153.
18. Ibid., 155–156.
19. These were the Todas, Badagas, Kurumbas, Irulas, and Kotas. The Todas, whom some of the earliest studies constructed as a lost Roman tribe, are perhaps one of the most studied groups in the subcontinent.
20. Metz, *Tribes Inhabiting the Neilgherry Hills*.
21. Ibid.
22. Ibid.
23. Ibid.
24. See Rosalind O'Hanlon and David Washbrook, "After Orientalism: Culture, Criticism and Politics in the Third World," *Comparative Studies in Society and*

History 34.1 (January 1992): 141–167; Gyan Prakash, "Can the Subaltern Ride? A Reply to O'Hanlon and Washbrook," *Comparative Studies in Society and History* 34.1 (1992): 168–184; Spivak, "Can the Subaltern Speak?"

25. As I have argued here, however, narratives of labor and agriculture are ideologically of a piece with scientific ideologies of progress.
26. As Gayatri Spivak has argued, literary and historiographical skills must be actively intertwined when we read colonial histories against the grain. Whereas the historian "unravels the text to assign a new position to the subaltern," the literary scholar "unravels the text to make visible the assignment of subject positions," she claims. See Gayatri Spivak, "A Literary Representation of the Subaltern," in *Subaltern Studies V: Writings on South Asian History and Society*, edited by Ranajit Guha (New Delhi and New York: Oxford University Press, 1987), 91. I defer, for now, the question of how we might reinscribe new subject positions in the agonistic domain of postcolonial science, while noting that the task of renarrating histories through reconstructed subaltern subjectivities will be a central programmatic issue in any attempt to formulate a cultural studies of colonial science.
27. Metz, *Tribes Inhabiting the Neilgherry Hills.*
28. Ibid.
29. Ibid.
30. Ibid.
31. Translated by and cited in M. B. Emeneau, *Toda Songs* (Oxford: Clarendon Press, 1971), 625.
32. Civilian, *The Civilian's South India*, 123–125.
33. Ibid., 132–133.
34. H. R. Morgan, *Forestry in Southern India*, edited by J. Shortt (Madras: Higginbotham, 1884).
35. Ibid.
36. Ibid.
37. Ibid.
38. Ibid.
39. Ibid.
40. Ibid.
41. Inspector-General Cole, *Madras Government Medical Department Report*, 1865 (Tamilnadu State Archives).
42. Ibid.
43. Ibid.
44. J. Shortt, *Handbook to Coffee Planting* (Madras: Government Press, 1864).
45. Captain John Ouchterlony, *Memoir to the Madras Government* (Madras: Government Press, 1847).
46. Ibid.
47. See Edward P. Thompson, *The Making of the English Working Class* (New York: Vintage Books, 1966); Raymond Williams, *The Country and the City* (New York: Oxford University Press, 1973); Raymond Williams, *Problems in Materialism and Culture: Selected Essays* (London: Verso, 1980).

Chapter 3 ***Forests***

1. "A Junglewallah's Letter," from "AWP" to his wife, Wynad, November 1878, in *Indian Forester* 4 (1878).
2. *1924 Review of Forest Administration in Madras Presidency.* Administrative Reports of Forests in Madras Presidency, 1921–1926, India Office Library Collection (IOR): V/24/1291 and V/24/1292.
3. *Report on Forest Administration, Madras Presidency, 1928–29* (Madras: Government Press), Tamilnadu State Archives (TNSA). Each state's regional forests were administered by an administrative superintendent called the forest conservator; they were all overseen by a chief conservator of forests for the state.
4. *Proceedings of Chief Conservator of Forests*, No. 545, October 1930. In *Board's Proceedings, Forest Series*, Chief Conservator of Forests (Madras: Government Press, 1930).
5. Ibid. *Taungya* was a system in which forest inhabitants were allowed to cultivate agriculture plots on forest lands (officially owned by the state), if they would plant and maintain trees on that plot in accordance with Forest Department stipulations. *Podu* was the practice of planting crops on patches of forest that had been felled and set alight. The ash served as a fertilizer, and the plot gave a high yield for two or three years; it is also known as slash-and-burn agriculture.
6. Administrative Report of Forests in Madras Presidency, 1907–1908, IOR: V/24/1289. It is possible that "drunkenness" was used as a veil from behind which to attack an otherwise unapproachable opponent.
7. "Influence Exerted by Trees on the Climate and Productiveness of the Peninsula of India," *Indian Forester* 4 (1878).
8. Following a famine in south India, a Court of Directors' Dispatch (No. 21 of 7 July 1874) requested the Government of India "to ascertain the effect of trees on the climate and productiveness of a county, and the results of extensive clearances of timber." Cited ibid.
9. "Influence Exerted by Trees." The reference is to Charles Darwin's *The Origin of Species by Means of Natural Selection, or The Preservation of Favoured Races in the Struggle for Life*, 3rd ed. (London: John Murray, 1861), 74–75.
10. "Influence Exerted by Trees."
11. E. P. Stebbing, "The Forests of India," British Association for the Advancement of Science paper, Edinburgh, 7–14 September 1921. Reprinted as "The Forests of India and the Development of the Indian Forest Department," in *Indian Forester* 48.2 (February 1922): 93.
12. Other well-known foresters (Richard Henry Beddome from 1870 and Campbell Walker from 1881) served as inspectors of forests, and in 1881 Dietrich Brandis (inspector-general of forests) commenced a two-year inspection of the Madras forests, which culminated in the passing of the Madras Forest Act in 1882. For a useful history of Madras forestry, see Hugh Cleghorn, *Manual of Administration for Madras Presidency* (Madras: Government Press, 1885), vol. 1; Cleghorn, *The Forests and Gardens of South India* (London: W. H. Allen, 1861); and *Proceedings of the Chief Conservator of Forests* 16 (January 1929) (Madras: Government Press, 1929).
13. See Stebbing, "The Forests of India," and H. R. Morgan, *Forestry in Southern*

India, edited by J. Shortt (Madras: Higginbotham, 1884). For an excellent overview of the Forest Acts, the debate over the 1882 Madras Forest Act, and consequences for local use, see Madhav Gadgil and Ramachandra Guha, *This Fissured Land: An Ecological History of India* (New Delhi and New York: Oxford University Press, 1992), 124ff.

14. See S. Ravi Rajan, "Scientific Forestry and the British Empire," D. Phil. thesis, Oxford University, 1994, 124. Dietrich Brandis, a German forester who through his marriage was closely connected to the governor-general of India, was appointed inspector-general of forests for India in 1864. With Hugh Cleghorn (conservator of forests for Madras from 1856), he laid the institutional foundation for scientific forestry in south India from the 1860s up to the passing of the Forest Act. Ibid., 128–129. See also Stebbing, "The Forests of India," 96–97. Stebbing reports the profits in the period 1874–1879 to be Rs. 1,500,000 and in 1919–1920 as Rs. 22,000,000. He considers other parts of the British Empire to be fifty years behind India in the organization of forestry, and recommends that the Indian model be applied to other countries.
15. *Report of the Proceedings of Forest Conference, Simla 1875*, edited by D. Brandis and A. Smythies (Calcutta: Government Press, 1876); original emphasis.
16. Ibid.
17. Ibid.
18. B. H. Baden-Powell, "The Political Value of Forest Conservancy," *Indian Forester* 2.3 (January 1877): 280–282.
19. Ibid., 282–283.
20. Ibid., 284, original emphasis.
21. *Administrative Report on Forests in Madras Presidency 1861–1862*, IOR: V/24/1281. The same report notes: "I had several applications for land, and it became necessary to draw a line, accordingly, the Teak belt was chosen. . . . In the north of Wynad, just outside the Teak belt, there is a good deal of Blackwood; but unless we could fell it, and were prepared to remove it, I do not see the use of keeping out the pioneers of civilization, viz. the coffee planters, for the sake of merely conserving timber we cannot make use of in any way." Here we can see that, although government and private interests were not always the same and indeed were sometimes competing, the foresters considered the general project of the private enterpreneurial planters as progressive, in that it ushered in modernity. I will return to the planters later in this chapter.

    Other demands on the land, such as mining, were encouraged. In the late nineteenth and early twentieth centuries, mining was explicitly encouraged by the Forest Department; leases were given for mica, manganese, diamond, and gold prospecting (see *Administrative Reports* for 1906–1907, 1909–1910, 1910–1911, IOR:V/24/1289). At the same time, tribes were strictly restricted from using the forest, and evicted from their departmentally owned plots if found committing "forest offences."
22. The tribals included 200 Kurumbas, 50 Kurichiyars, 100 Panniyars and Pulayars, 50 "Chettys and Squatters," according to the *Administrative Reports of Forests in Madras Presidency, 1861–1862*, IOR: V/24/1281.

23. *Administrative Report of Forests in Madras Presidency, 1866–1867*, IOR: V/ 24/1281.
24. Board's Proceedings, Public (Political), 14 Nov 1938, no. 1869 (TNSA), 1.
25. Ibid.
26. *Administrative Report of Forests in Madras Presidency, 1907–1908*, IOR: V/ 24/1289.
27. Revenue Department Memo no. 2617 E/ 15–11, 8 December 1916 (TNSA).
28. Ibid.
29. Ibid.
30. *Administrative Report of Forests in Madras Presidency, 1927–1928*, IOR: V/ 24/1293.
31. The idea of a *normalbaum* came from the principles of German scientific forestry. See William Schlich, *A Manual of Forestry*, 2d ed., 5 vols. (London: Bradbury, Agnew, 1896), 1:16–17.
32. *Working Plan for the Mudumalai Forests, 1927* (TNSA), 26.
33. At this time there were also schools for Sholagas in north Coimbatore, Mulcers in Mount Stuart, Kurumbas in Mudumalai, and Koyas in Bhadrachalam taluk. *Selections from the Records of Madras Government, Revenue Department* (Forest Series), Rout. N. 100, 21 January 1921. The following material is based on this report unless otherwise stated.
34. Ibid.
35. Ibid.
36. Ibid.
37. *Working Plan for the Deciduous Forests of the Wynad Plateau* (Madras: Government Press, 1929), 16.
38. Ibid., 43.
39. Ibid., passim.
40. Ibid., 11.
41. Ibid., 13–14,
42. Ibid., 15.
43. Memo from District Forest Officer Coimbatore to Chief Conservator of Forests, Madras, *Selections from the Records of Madras Government, Revenue Department* (Forest Series), Madras, 17 July 1940, re: Labour Supply (TNSA).
44. See, for example, Gadgil and Guha, *This Fissured Land*, and Vandana Shiva, *Staying Alive: Women, Ecology and Survival* (London: Zed Books, 1988).
45. I focus here on the instances of recorded intransigence on the part of tribal groups during Crown rule. I do not go into the reluctance and protest on the part of kings and landowners to part with their land, although there is much evidence of this.
46. There are, however, some recorded instances of petitions on behalf of village communities. I cite one from 1935 later in this section.
47. *Administrative Report on Forests in Madras Presidency, 1921–1922*, IOR: V/ 24/1291.
48. See K. N. Panikkar, *Against Lord and State: Religion and Peasant Uprisings in Malabar, 1836–1921* (New Delhi and New York: Oxford University Press, 1989);

R. H. Hitchcock, *Peasant Revolt in Malabar* (New Delhi: Usha Publications, 1983).

49. *Administrative Report on Forests in Madras Presidency, 1921–1922*, IOR: V/24/1291.
50. The Malayalees, or Malialis, were tribal groups that lived in Salem District. They are not to be confused with the Malayalam-speaking residents of Kerala. The tribal group name, Malayalee, appears to have its roots in *mala* (hill) and *ál* (person).
51. *Administrative Report on Forests in Madras Presidency, 1920–1921*, IOR: V/24/1291.
52. Ibid.
53. Letter from Doris Hitchcock, Scottish missionary, to her mother, 6 October 1921; private collection. I thank Rajan Chetsingh for sharing this personal correspondence with me.
54. Procedings of Chief Conservator of Forests, 1924. In *Administrative Reports of Forests in Madras Presidency 1921–1926*, IOR: V/24/1292.
55. Letter from Private Secretary to H. E. Governor to the Secretary, Development Department. D.O. 1–305/35, 27 September 1935 (TNSA).
56. *Administrative Report on Forests in Madras Presidency, 1924–1925*, IOR: V/24/1292. The chief conservator of forests also notes here that the aims of forest administration are threefold: (a) to obtain the highest possible income from that part of the forest estate which is commercially valuable, (b) to protect forest growth, and (c) to preserve small forests for local use. He complains that the last made the department unpopular and fetched no revenue; he praises the new decision to transfer them to the Revenue Department, after which the Forest Department continues the management of tribal settlements (as no other department has the infrastructure to manage them), although the Revenue Department pays for it.
57. *Annual Report on Forest Engineering Branch* (Proceedings, Forest Series, Chief Conservator of Forests, 1924–1925) (TNSA).
58. Ibid.
59. *Annual Report of Forests in Madras Presidency, 1924–1925*, IOR: V/24/1292.
60. See the chapter, "Representing Authority in Victorian India," in Bernard S. Cohn, *An Anthropologist among the Historians and Other Essays* (Delhi and New York: Oxford University Press, 1987), 632–682, for an analysis of the spectacularization of authority in a durbar celebration.
61. *Proceedings of Chief Conservator of Forests, 1933* (Administrative Report of Forests in Madras Presidency, 1933–1934), IOR: V/24/1295.
62. Ibid.
63. *Proceedings of Chief Conservator of Forests, 1937* (Administrative Report of Forests in Madras Presidency, 1936–1937), IOR: V/24/1296.
64. *Annual Report of Forests in Madras Presidency, 1924–1925*, IOR: V/24/1292.
65. Historians who would point to failure include Anson Rabinbach, *The Human Motor: Energy, Fatigue, and the Origins of Modernity* (New York: Basic Books, 1990); and Daniel R. Headrick, *The Tentacles of Progress: Technology Transfer in the Age of Imperialism, 1850–1940* (New York: Oxford University Press, 1988).

66. See, for example, almost any *New York Times* article on Third World development today; for example, "India's 5 Decades of Progress and Pain," John F. Burns, *New York Times*, 14 August 1997; "India Ruefully Takes Stock of 49 Years," John F. Burns, *New York Times*, 16 August 1996; for the most common Indian version of this argument, see Pushpa M. Bhargava and Chandana Chakravarty, "Of India, Indians, and Science," *Daedalus* 118.4, special issue on "Another India" (Fall 1989): 353–368.

CHAPTER 4 ***Plantations***

1. P. H. Daniel, *Red Tea* (Madras: Higginbotham, 1969), 76.
2. Ranajit Das Gupta, "Plantation Labour in Colonial India," *Journal of Peasant Studies*, Special Issue on "Plantations, Proletarians, and Peasants in Colonial Asia," 19.3–4 (1992): 173–198. For a planter's account, see S. G. Speer, ed., *UPASI: 1893–1953*, Diamond Jubilee Publication (Coonoor: United Planters' Association of Southern India [UPASI], 1953), 158–173. On debt bondage, see Jan Bremen and E. Valentine Daniel, "The Making of a Coolie," *Journal of Peasant Studies* 19.3–4 (1992): 268–295.
3. For an overview of modernization theory, see M. C. Howard and J. E. King, *A History of Marxian Economics*, vol. 2, *1929–1990* (Princeton: Princeton University Press, 1992).
4. Das Gupta, "Plantation Labour in Colonial India," 189.
5. Ibid., 173, 186, and 193.
6. Daniel worked as a medical officer and union organizer in plantations before and after Independence. Although *Red Tea* is presented in the form of a novel, Daniel states in his introduction that the conditions represented in it are based on his actual findings, of which he tries to "present a correct and true picture." As the novel is written with an explicitly documentary purpose, and its context is described as the conditions leading up to and resulting from the Madras Planters Act of 1903, I use it as a documentary source for this period rather than treating it as only a piece of creative fiction.
7. Daniel, *Red Tea*, 65. He lays out the details of the indebting process, in a description that fits with Das Gupta's and other plantation historians' findings. Daniel also records that sexual harassment of women workers was a common practice.
8. *Planting Opinion*, 10 December 1896. UPASI Headquarters, Coonoor.
9. UPASI Scientific Department, "Indian Labour in the F.M.S." *Planters' Chronicle* 15.35 (August 1920): 642.
10. Rudolph D. Anstead in S. Playne, *Southern India: Its History, People, Commerce, and Industrial Resources*, compiled by Somerset Playne, assisted by J. W. Bond, edited by Arnold Wright (London: Foreign and Colonial Compiling and Publishing, 1914–1915). "Kelly" might be a misprint, as I assume the writer is referring to the Ketti Valley estates, 258.
11. In *The Human Motor*, Rabinbach reminds us that "at the beginning of the nineteenth century, ethnographers attempted to demonstrate that civilization not only promoted the habit of industry, but encouraged physical development as well";

Anson Rabinbach, *The Human Motor: Energy, Fatigue, and the Origins of Modernity* (New York: Basic Books, 1990), 30. Further, he notes that climate, class, and idleness were also considered to be closely connected in the middle of the nineteenth century. "The virtues of order, thrift and industry were set against the ravages of idleness, laziness, vagabondage and the dissolute life of crime. The association of 'laboring classes and dangerous classes' that belongs to this epoch is rooted in this perception that workers with too much free time turn inevitably to drink and crime. The literature on work written by middle-class reformers is edifying and uplifting, directed against the debilitating effects of idleness. Work was . . . prophylaxis against the dissolute qualities of inordinate leisure and, of course, against the concomitant effects of drink" (ibid., 30–31).

12. Ibid., 35.
13. Ibid., 4–5, 8.
14. "The Cooly and Drink," *Planters Chronicle* 15.36 (1920): 601–602. Reprinted from *Malayan Tea and Rubber Journal.*
15. Rabinbach also sees the shift from labor/idleness to labor power/energy as crucial to the development and legitimation of liberal social thought in modern Europe.
16. Ross was a doctor, well known for his discovery of the link between the anopheles mosquito and the transmission of malaria. A major part of the research leading up to this discovery had been done in the Nilgiris. The UPASI and Ross collaborated on several projects.
17. "International Congress of Tropical Agriculture," *Planters Chronicle* 9 (July 1914): 438.
18. "Immigrants," *Planters' Chronicle* 9 (2 August 1914): 257–258.
19. *Planting Opinion*, 12 August 1899.
20. Daniel R. Headrick, *The Tentacles of Progress: Technology Transfer in the Age of Imperialism, 1850–1940* (New York: Oxford University Press, 1988), 209.
21. Ibid., 210–211.
22. Ibid., 16.
23. For a revealing representation of the relationship between foresters, natives, animals, and the Indian jungle, see Rudyard Kipling, "In the Rukh," in his *Many Inventions* (London and New York: Macmillan, 1893).
24. Quoted in Headrick, *Tentacles of Progress*, 210.
25. Although planters were isolated from the rest of the country, they quickly made oases of "Englishness" for themselves. Nature was both discursively and materially altered in accordance with English desires and values. Chapter 2 explored some "English" constructions of the Nilgiris. Here I focus on the moral discourse of progress within colonizer-native relations.
26. *Times of Ceylon*, 10 May 1902, report on Chief Commissioner Cotton's inquiry into tea-plantation labor in south India.
27. From the *London Daily Mail*, reprinted in *Planting Opinion*, 12 August 1899.
28. *Planting Opinon*, 12 August 1899.
29. Reported in *Planters' Chronicle* 62.16 (15 August 1947).
30. *Planters Chronicle* 15 (January 1920).
31. A Government of India resolution on 25 March 1896 appointed a committee

"to conduct an enquiry into certain matters forming the subject of a representation made to His Excellency the Viceroy by the United Planters' Association of Southern India at Madras in December 1985." Cited in Speer, *UPASI: 1893–1953*, 6.

32. Daniel, *Red Tea*, 1.
33. *Planters Chronicle* 9 (1914): 88. See also the report on the Third International Congress of Tropical Agriculture, held at the Imperial Institute, ibid., 436–438.
34. In 1914 the *Quarterly Journal of the Scientific Department of the Indian Teas Association* reported that about 10 lbs per acre were being lost each year because of chronic fungus disease alone, and over 200,000 pounds sterling lost in epidemic diseases (*Planters' Chronicle* 9 (February 1914): 89). See also *Madras District Gazeteer: Malabar 1933* (Madras: Government Press) (Tamil Nadu State Archives, Chennai).
35. Cecil Wood, "The Farmer and the Expert," *Planters' Chronicle* 15 (1920): 61.
36. *Planters' Chronicle* 24 (1929), report of speech by the Rt. Hon. Lord Melchett, "India as Supplier of Empire Resources," 1157.
37. Ibid., 1157–1158.
38. Ibid., 1158–1159.
39. T. J. Barron, "Science and the Nineteenth-Century Ceylon Coffee Planters," *Journal of Imperial and Commonwealth History* 16.1 (1987): 5, emphasis added.
40. Ibid., 5, 7, emphasis added.
41. Ibid., 9, emphasis added.

Chapter 5 *Ethnographers*

1. Report of a correspondent of *A Friend of India* from Kalastri, N. Arcot, 16 March 1878. Cited in William Digby, *The Famine Campaign in Southern India, 1876–1878* (London, 1878), found in Connemara Public Library, Chennai.
2. See, for example, James Clifford and George Marcus, eds., *Writing Culture: The Poetics and Politics of Ethnography* (Berkeley and Los Angeles: University of California Press, 1986); James Clifford, *The Predicament of Culture: Twentieth-Century Ethnography, Literature, and Art* (Cambridge: Harvard University Press, 1988); and Kamala Visweswaran, *Fictions of Feminist Ethnography* (Minneapolis: University of Minnesota Press, 1994).
3. As Johannes Fabian writes, "The critique of anthropology is too easily taken for moral condemnation. But at least the more clearheaded radical critics know that bad intentions alone do not invalidate knowledge. For that to happen it takes bad epistemology which advances cognitive interests without regard for their ideological presuppositions." Johannes Fabian, *Time and the Other: How Anthropology Makes Its Object* (New York: Columbia University Press, 1983), 33.
4. See A. C. Haddon, "The Teaching of Ethnology at Cambridge: Lectures to Missionaries," *Manchester Guardian*, 15 January 1903; W.H.R. Rivers, "The Government of Subject Peoples," in *Science and the Nation: Essays by Cambridge Graduates*, edited by W.H.R. Rivers (New York: Books for Libraries Press, 1917); George Henry Pitt-Rivers, *The Clash of Culture and the Contact of Races: An Anthropological and Psychological Study of the Laws of Racial Adaptability,*

with Special Reference to the Depopulation of the Pacific and the Government of Subject Races (London: Routledge and Sons, 1927).

5. George Stocking and Henrika Kuklick have provided comprehensive and important surveys of British anthropology. See George Stocking, *Victorian Anthropology* (New York: Free Press, 1987); and Henrika Kuklick, *The Savage Within: The Social History of British Anthropology, 1885–1945* (Cambridge: Cambridge University Press, 1991). For other histories, see Talal Asad, ed., *Anthropology and the Colonial Encounter* (London: Ithaca Press, 1973); Christopher Herbert, *Culture and Anomie: Ethnographic Imagination in the Nineteenth Century* (Chicago: University of Chicago Press, 1991); and Marvin Harris, *The Rise of Anthropological Theory: A History of Theories of Culture* (New York: Crowell, 1968).
6. Anthropology might be said to have been formally recognized as a discipline in 1884, with the formation of the British Association for the Advancement of Science Section H, devoted to the study of anthropology; and the creation of its first university post, when E. B. Tylor was given a readership in anthropology at Oxford. See Kuklick, *The Savage Within*, chapter 5. For a history of the struggle between "ethnography" and "anthropology," see Kuklick, *The Savage Within*, and Stocking, *Victorian Anthropology*. In this chapter, I review the history of the discipline of anthropology, acknowledging that this term won the disciplinary wars of naming. I refer to the characters in my history as ethnographers, however, sticking to their self-descriptions and the usage of the time.
7. Kuklick, *The Savage Within*.
8. Ibid., 44.
9. Ibid., 182.
10. Ibid., 184.
11. Ibid., 188.
12. Ibid., 189.
13. Ibid., 240.
14. Charles Morrison, "Three Styles of Imperial Ethnography: British Officials as Anthropologists in India," in *Knowledge and Society: Studies in the Sociology of Culture Past and Present, A Research Annual*, vol. 5, edited by Henrika Kuklick and Elizabeth Long (Greenwich, Conn.: JAI Press, 1984), 143.
15. George Stocking, "The Nineteenth-Century Concept of Race," lecture delivered to the Department of History, UCLA, 3 March 1969, quoted with kind permission of the author.
16. He is referring to Topinard's "list of statures," and a tabulation that excludes tribes of Central Africa and Morocco.
17. Edgar Thurston, "Notes on the People of Malabar 1900," with sections by Fred Fawcett and Florence Evans, *Madras Government Museum Bulletin* 3.1 (1900): 1–85.
18. With the possible exception of the Syrian Christians. With their claims of Brahminical origins and apostolic Christian traditions, their access to education was doubly privileged. See Susan Visvanathan, *The Kerala Christians* (New Delhi: Oxord University Press, 1993); and Susan Kauffman, "The Syrian Christians of Malabar," D. Phil. thesis, Cambridge University, 1979.

19. See the following studies by L. K. Ananthakrishna Iyer: *The Cochin Tribes and Castes*, 2 vols. (London: Higginbotham, and London: Luzac, 1909–1912); "The Mala Arayans or Kanikkars of the Travancore Forests," *Man in India* 2 (1922): 55–61; "The Malsers of the Cochin Forests," *Man in India* 3 (1923): 36–58; "Inheritance among Primitive Peoples," *Man in India* 6 (1926): 194–196; *The Mysore Tribes and Castes*, 4 vols. (Mysore: Mysore University, 1928–1935); "The Kadu-Kurumbas," *Man in India* 9 (1929): 223–229.
20. L. P. Vidyarthi and B. K. Rai, *The Tribal Culture of India* (Delhi: Concept Publishers, 1977), 3–4. This textbook is used as a standard graduate anthropology text in Kerala. I thank K. Seethu for sharing her textbooks with me.
21. The Todas reportedly gave an account of themselves to anthropologists as having descended from Ravana. This genealogy may have been suggested to them during some prior contact with upper-caste groups. See Edgar Thurston, *Ethnographic Notes in Southern India* (Madras: Government Press, 1906).
22. Vidyarthi and Rai, *The Tribal Culture of India*, 15.
23. Johannes Fabian's radical critique of the epistemological and political constructions of time in anthropological thought is a model by which we could combine a critique of both colonial and indigenous anthropology. Fabian suggests that the relegation of "primitive" people to the past of "civilized" societies is part of a distancing device central to ethnographic methodology. He refers to this device as the "denial of coevalness," which he defines as "a persistent and systematic tendency to place the referent(s) of anthropology in a Time other than the present of the producer of anthropological discourse" (Fabian, *Time and the Other*, 31).
24. Captain Henry Harkness, *Description of Singular Aboriginal Race Inhabiting the Summit of the Neilgherry Hills or Blue Mountains of Coimbatore in the Southern Peninsula of India* (London: Smith, Elder, 1832), 85–86, original emphasis.
25. Ibid., 88–89.
26. W.H.R. Rivers, *The Todas* (London: Macmillan, 1906), 716–718. See pp. 2–3 for Rivers's account of why his initial contact with the Todas was so compelling that he "gave up the intention of working with several different tribes, and devoted the whole of [his] time to the Todas."
27. Edgar Thurston assisted by K. Rangachari, *Tribes and Castes of Southern India* (Madras: Government Press, 1909). Although Rangachari is explicitly thanked in Thurston's introduction, his name never joined the ranks of generally acknowledged great anthropologists, and even in Vidyarthi's 1977 textbook, which lists every major and minor anthropologist, he is only referred to as Thurston's assistant. Kuklick observes that colonial political officers' duties were often represented in terms very similar to the field work of functionalist anthropologists: "The rural colonial official was supposed to travel frequently among his subjects with no European companions, making personal contact with as many of his charges as possible. In truth, he toured with a retinue of local people—bearers, guides, servants and interpreters—but these were nearly invisible in his official reports and memoirs" (Kuklick, *The Savage Within*, 189–190).
28. Edgar Thurston, *The Madras Presidency with Mysore, Coorg and the Associated*

States. Provincial Geographies of India series, edited by T. H. Holland (Cambridge: Cambridge University Press, 1913), 1.

29. Editor's preface, ibid., 5–6.
30. Morrison, "Three Styles of Imperial Ethnography."
31. Kuklick and Long, *Knowledge and Society*, vol. 5; Morrison, "Three Styles of Imperial Ethnography."
32. Marshall, *A Phrenologist amongst the Todas, or the Study of a Primitive Tribe in South India, Their History, Character, Customs, Religion, Polyandry, Language* (London: Longmans, Green, 1873).
33. La Peyrere's *Praeadamitae* (1655), cited in Stocking, *Victorian Anthropology*, 12. See also Margaret T. Hodgen, *Early Anthropology in the Sixteenth and Seventeenth Centuries* (Philadelphia: University of Pennsylvania Press, 1964).
34. Stocking, *Victorian Anthropology*, 12, 44.
35. In *Victorian Anthropology*, Stocking sets the stage for understanding colonial anthropology. On biblical anthropology, he comments, "From a very broad historical perspective, in which the history of anthropological thought may be seen as an alternating dominance of the biblical and the developmental traditions, the pre-Darwinian period in Britain is one in which, after a century in retreat, the biblical tradition reassumed a kind of paradigmatic status" (44). In *A Phrenologist amongst the Todas*, Marshall's arguments, including the use of comparison with American Indians, follow the common tropes of the biblical model, which assumed human degeneration from an original innocent state.
36. G .W. Blair, *Station and Camp Life in the Bheel Country: A Brief History of the Origin, Aims and Operations of the Irish Presbyterian Jungle Tribes' Mission, with an Account of the Bheels* (Belfast, 1906), 1, IOR: T 47278.
37. Ibid.
38. Reverend Samuel Mateer, *Native Life in Travancore* (London: London Missionary Society, 1883), 64.
39. Marshall, *A Phrenologist amongst the Todas*, 265.
40. Ibid., 266.
41. Ibid., 269–270.
42. Ibid., 266.
43. Each natural formation had deities or stories associated with it, and the Todas gave thanks to these deities before using resources such as water and wood. Mukurti Peak, in particular, was held to be the one of most sacred natural formations, as it was considered the place closest to Amnor, the next world, and the place from where souls of dead Todas and their buffaloes leapt over from this world when they died. Note that Marshall describes Mukurti Peak in his nature rhapsody. He goes on to claim that the Todas could not have chosen this spot as sacred for its beauty, for their skulls show no organs of taste. They must merely have noted "that Mukurti is the highest and western-most hill, nearest the setting sun, and therefore the most suitable locality for the purpose" (ibid., 267; original emphasis).
44. I first introduced it in the context of the poem on "Nilgiri Sunshine," arguing that we could see a double binary, the high nature/high culture pair along with the low nature/low culture pair. The former was characterized by a firm separa-

tion between nature and culture, while the latter was characterized by a close relationship between them, such that "primitive culture" was almost inseparable from "brute nature."

45. Arrack or charayam is country liquor. Mateer, *Native Life in Travancore*, 69–70.
46. Ibid., 70. Though the Kanikars were counted in the census as a low caste, they considered themselves higher than the Shanars and Izhavas, and might have adopted the beef taboo as part of this identity.
47. Ibid.
48. For a similar analysis of British missionary encounters with South African natives, see Jean Comaroff and John Comaroff, *Of Revelation and Revolution: Christianity, Colonialism, and Consciousness in South Africa* (Chicago: University of Chicago Press, 1991).
49. In one sense, the reorganization of the face of the forests was spectacularly obvious: clear-felling; the establishment of departmental fire lines and plantation areas; the planting, protecting, and harvesting of monocultures, and so on, displayed the power of colonial foresters to literally change the face of the land. To local inhabitants of these forests, these changes were hard to miss, as they dramatically reorganized their physical surroundings. A work of fiction that powerfully captures the myriad levels at which such a change might be experienced is Ursula LeGuin's novella, *The Word for World Is Forest* (New York: Berkeley Books, 1976).
50. Mateer is reporting the narrative offered to him in a conversation with a planter, Mr. Emlyn. See Mateer, *Native Life in Travancore*, 70–71.
51. L. P. Vidyarthi, *The Rise of Anthropology in India: A Social Science Orientation* (New Delhi: Concept Publishers, 1978), 294. Vidyarthi mentions Samuel Mateer (whom he refers to as "Samul Mati") on p. 278.
52. See Steven Shapin, "Homo Phrenologicus: Anthropological Perspectives on an Historical Problem," in *Natural Order: Historical Studies of Scientific Culture*, edited by Barry Barnes and Steven Shapin (London: Sage, 1979).
53. *Report of Madras School of Industrial Arts* (Madras: Government Press, 1853). IOR Tracts, vol. 144.
54. "Various mineral products of this Presidency have been brought to bear in the manufactures effected; such as porcelain earths, and clays, gypsum, quartz, felspar, granites, steatite, corundum, magnesite, galena, manganese, plumbage, emery, . . . antimony, . . . the ores of iron, and colored earths. While these and other products, either brought to light, or discovered in new localities and in greater variety, by the same means, are likely to become articles of large and profitable export." Ibid.
55. Letter dated 29 July 1862 from V. Kistnamachari, Deputy Inspector of Schools, Madras Presidency, to Alexander Hunter, Superintendent, Madras School of Industrial Arts. IOR Tracts, vol. 144.
56. Ibid.
57. Ibid.
58. *The Quarterly Report of the School*, 15 November 1866, reports that photographs were taken of the Nilgiri tribes (whom they identify as "Todas, Burghers, Kotahs, Irulers, Karumbers, . . . and Cassaveros"); IOR Tracts, vol. 144. Photographic

documentation of tribal faces and bodies was considered valuable ethnographic data, especially since many of their "inherent" characteristics were believed to be readable from visages. Photographs were also taken of the converts of the Free Church Mission schools of Madras, which, like the other missions I have mentioned, recruited converts among tribals and lower castes.

59. These castings are reported in the *Quarterly Report* for 15 April 1867, among reports of the making of vases, tiles, pottery, tools, and chemical dyes. They were sent to the Paris Exhibition, the Royal Society of Arts in London, and the Schools of Art in Calcutta, Bombay, Jaipur, and Jabalpur.
60. *Report of Madras School of Industrial Arts*, May 1868. IOR Tracts, vol. 144.
61. The remainder of this paragraph is quoted in this chapter's epigraph. The extract is from Digby, *The Famine Campaign in Southern India 1876–78*, emphasis added. The words are not Digby's, however, but are quoted from a local newspaper article on the famine. Digby nevertheless quotes the same correspondent extensively for information on the famine. The discussion above is offered without critique, and presented as firsthand information on famine conditions. Note how the correspondent, in turn, quotes a geography text (with ethnographic information) in order to legitimate and naturalize a social and moral choice.
62. Josiah Waters Coombes, *The Juvenile Criminal in Southern India* (Madras: Lawrence Asylum Press, 1908), 129. The data collected is with special reference to the Chinglepet Reformatory, just outside Madras city. In Madras Presidency at this time, criminal reformations were placed under the Education Department rather then the Police Department.
63. Ibid., Plate 33.
64. Ibid., 6–7.
65. Quoted in Nicholas Dirks, "The Policing of Tradition: Colonialism and Anthropology in Southern India," manuscript, c. 1991, 22.
66. Ibid., 22, 29.
67. Quoted by Dirks, ibid., 28–29, from G.O. 1838/9–9/93/Judicial. Anthropometry as a policing tool gradually yielded to fingerprinting, which had been developed in the 1850s in Bengal, and which by the end of the century was more sophisticated than the former. However, anthropometry did not lose its appeal among scientists, and among the committed band of followers it had gained among police officials. In addition, Risley's mid-century call for a nationwide effort to advance anthropological knowledge resulted in a spate of studies on his model from the end of the nineteenth century through the first decades of the twentieth century; many were inspired by Risley's model to pursue the physical and quantitative aspects of anthropology. There is still much need for a critique of anthropometry within India, as a survey of mid-twentieth-century Indian anthropology would show. Such well-known nationalist anthropologists as Iravati Karve have published anthropometric monographs on the tribes of India, incorporating within the old paradigms up-to-date techniques such as blood analysis.
68. Christopher Pinney, "Colonial Anthropology in the Laboratory of Mankind," in *The Raj: India and the British 1600–1947*, edited by C. A. Bayly (London: National Portrait Gallery, 1990), 287.
69. The Criminal Tribes Act (Act XXVII of 1871) was initially only enforced in

the United Provinces, Punjab, Bengal, Awadh, and the North West Provinces. In 1883 an enquiry was conducted to assess the need for extending it to all of India; the response was overwhelmingly positive from Madras Presidency, among the police and other administrative officials surveyed; see *Selections from the Records of the the Government of India Home Department*, No. CCC, Calcutta, 1883; Records Office, Kozhikode, Kerala. The Criminal Tribes Act was subsequently (1911) extended to all of India, and successfully applied to several Madras Presidency tribes, including the Chenchus and the Yenadis who, as we have seen, had long been viewed as problems by the Forest Department, missionaries, and educationists.

70. Muhammed Abdul Ghani, *Notes on the Criminal Tribes of the Madras Presidency* (Madras: Government Press, 1915) (TNSA).
71. This is an extract from the Madras Criminal Tribes Manual 1935, which includes the wording of the Criminal Tribes Act, with amendments up to 1935.
72. C. W. Gayer, *Lectures on Some Criminal Tribes of India and Religious Mendicants* (Madras: Government Press, 1910). He quotes a speech by Mr. Chitnavis, CIE, Provincial Member, Central Province, during the budget debate of 1901. "There is yet another serious difficulty in the way of Indian industrial development. There is a growing dearth of labour at industrial centres, and more than one industry suffers in consequence. Appreciable relief can be afforded by Government in this matter by encouraging immigration of surplus population, . . . and indicting beggary except in the case of the aged the infirm and the disabled. . . . I think the time has come when the strong arm of the law should intervene to arrest its [beggary's] further progress . . . so that all strong-built and able bodied men who are now living on begging may be utilised for [agricultural] work; this may I think relieve scarcity of labour to a certain extent."
73. Ibid.
74. Ibid.
75. Summarized from Gayer, ibid.
76. Ibid.
77. Ibid.
78. Abdul Ghani, *Notes on the Criminal Tribes of the Madras Presidency.*
79. *Note Showing the Progress Made in the Settlement of Criminal Tribes of Madras Presidency up to January 1925* (Madras: Government Press, 1925) (TNSA).
80. Ibid.
81. This assertion is supported by studies in the history of colonial anthropology; see, for example, Pinney, "Colonial Anthropology in the Laboratory of Mankind," and essays in George Stocking, ed., *Colonial Situations* (Chicago: University of Chicago Press, 1991).
82. James Mill, *The History of British India*, 2d ed., 6 vols. (London: Baldwin, Craddock and Joy, 1820), vol. 1, 247.
83. See, for example, Thomas Kuhn, *The Structure of Scientific Revolutions*, 2nd ed. (Chicago: University of Chicago Press, 1970); and Richard Boyd, Philip Gasper, and J. D. Trout, eds., *The Philosophy of Science* (Cambridge: MIT Press, 1991).
84. Satya Mohanty, "Drawing the Color Lilne: Kipling and the Culture of Colonial

Rule," in *The Bounds of Race*, edited by Dominick LaCapra (Ithaca: Cornell University Press, 1991), 340.

85. For example, in England, Sir George Cayley, founder of the Polytechnic Institution, advocated social reform through popular technical education. Henry Mayhew gave a regular series of Monday night lectures in the summer of 1854, geared specifically to the "working classes," on "the CURIOSITIES of LIFE among the LABOURERS and POOR of LONDON." Cited in Richard Altick, *The Shows of London* (Cambridge: Belknap Press, 1978), 387. A 1936 study of Indian museums conveys a very similar picture. It dates the first museum to 1840, started in Calcutta by the Asiatic Society, devoted to economic geology. Madras followed, in 1846, when the government, "with the object of fostering scientific enquiries and pursuits," took over the collections of the Madras Literary Society (a branch of the Asiatic Society), and established a "Central Museum" at Fort St. George, which oversaw several regional museums. Describing "family visits" to the museum, the authors write: "It irresistibly reminded one of the old-fashioned fairs in England, with people walking round every one of the sideshows, and being rather amused. . . . They crowded round the cases showing indigenous games, village industries, agricultural operations, etc., . . . their faces wreathed in smiles. . . . It is doubtful whether reference was made to a label or could have been made. Nevertheless it was impossible not to feel that these visitors benefited from their visit. . . . To these visitors, the museum is a peep-show, a wonder house, a mansion full of strange things and queer animals, and the main appeal is to the Indian sense of wonder and credulity. . . . At Madras [on a festival day] 130,000 people trooped through the Museum. . . . And although the interior of the Museum was packed and produced the inevitable accompaniment of fingered glass, betel-nut spit and dirty marks on the walls, one realized that this was one of the surest ways of interpreting the outside world to the masses of India." S. F. Markham and H. Hargreaves, *The Museums of India* (London: The Museums Association, 1936), 61.
86. Tony Bennett, "The Exhibitionary Complex," *New Formations* 4 (Spring 1988): 87. Although Bennett is speaking specifically about scientific exhibitions in the nineteenth century, these disciplines shaped representation in the general social context of the time. The importance of the disciplines of geography, botany, forestry, and anthropology to colonial power should be evident by now.
87. Mill, *History of British India*, 1.12, 27. "Indian" is conflated, for the most part, with "Hindu," except for a comparative chapter on the "Mohammedans."
88. Ibid., 32.
89. The influence, of course, traveled the other way as well: scientific theories were influenced by cultural discourse.
90. Friedrich Engels, *The Condition of the Working Class in England in 1844–45* in *Karl Marx and Friedrich Engels: Collected Works*, translated by Richard Dixon et al., vol. 4 (New York: International Publishers, 1975), 390.
91. Roger Cooter notes, "Luminaries of early Victorian culture, such as Samuel Coleridge, John Stuart Mill, Harriet Martineau, Samuel Smiles, George Henry Lewis, Auguste Comte, and Henri de Saint Simon, became in some respects phrenology's intellectual guardians." Roger Cooter, *The Cultural Meaning of*

Popular Science: Phrenology and the Organization of Consent in Nineteenth-Century Britain (New York: Cambridge University Press, 1984), 7.

92. Karl Marx and Frederick Engels, *On Colonialism*, 4th ed. (Moscow: Progress Publishers, 1969), 80. This volume reprints the essays from the *New York Tribune* and other journalistic writings of Marx and Engels.
93. Ibid., 81.
94. Ibid., 87.
95. Ibid., 82.
96. Daniel R. Headrick, *The Tentacles of Progress: Technology Transfer in the Age of Imperialism, 1850–1940* (New York: Oxford University Press, 1988).
97. L. A. Krishna Iyer, *The Travancore Tribes and Castes*, 3 vols. (Trivandrum: Government Press, 1937–1941); *Social History of Kerala* (Madras: Book Centre Publications, 1968–1978).
98. "Indian Physical Anthropology and Raciology: Ramaprasad Chanda's Contribution," *Science and Culture*, 8.5 (November 1942): 201.
99. "Ramaprasad Chanda," *Science and Culture* 8.2 (August 1942): 67, original emphasis.
100. "Indian Physical Anthropology and Raciology," 207–208.
101. Ibid., 207.
102. *Dasyus* or *dasas* were believed by nineteenth-century anthropologists to be the Indian subcontinent's original indigenous inhabitants, dark-skinned opponents of the Vedic Aryans.
103. "Indian Physical Anthropology and Raciology," 204.
104. Ibid.
105. Sundaresa Iyer, *How to Evolve a White Race* (Madras, 1934) (TNSA). Iyer was an attorney by profession, which means that he had been a successful student either within the colonial educational system (whose ideological constitution is explored in by Gauri Viswanathan in *Masks of Conquest: Literary Study and British Rule in India* [New York: Columbia University Press, 1989]), or in England. His acquaintance with mathematical genetics indicates that he was probably influenced by the growing literature on eugenics at this time. Thus his model of race is not the same as the pregenetic nineteenth-century concept of race described by George Stocking (Stocking, "Nineteenth-Century Concept of Race"), which I discussed at the beginning of this chapter. Nevertheless, Iyer combines his genetic model of inheritance with a Lamarckian belief in the possibiity of the inheritance of acquired characteristics.
106. Ibid., 92.
107. Ibid., 51.
108. Ibid., 1.
109. Ibid., 52.
110. *Note Showing the Progress Made in the Settlement of Criminal Tribes*.
111. Quoted in Charles Gover, *The Folk Songs of Southern India* (Madras: Higginbotham, 1871), 81–87. Gover was a member of the Royal Asiatic Society and fellow of the Anthropological Society; he cites Metz's two-volume collection of Nilgiri songs and proverbs published by the Basel Mission Press. Once again (as with W. E. Marshall), we see Metz as the source of ethnographic data

for an anthropologist. Gover records these as Badaga songs; however, he also claims that the Todas have no songs, which is not true (French anthropologist Emeneau recorded large numbers of traditional Toda songs in the 1930s), so it is possible he confused the Badagas and the Todas. Gover comments: "The tribes are rapidly dying out under that strange law which will not permit a ruder tribe to coexist with modern civilization. They are kindly treated and are permitted to enjoy their simple holdings, but yet decrease in number year by year. Strong drink destroys the men. The women steadily deteriorate in contact with the Europeans" (99–100).

112. Ibid, 85, 73.
113. M. B. Emeneau, *Toda Songs* (Oxford: Clarendon Press, 1971), 516. At the time of this publication, Emeneau was professor of Sanskrit and linguistics at the University of California Berkeley. He first carried out his field work in the Nilgiris in 1936.
114. Ibid., 528–529.

Chapter 6 ***Christianity***

1. Address to the Basel Mission Society by "an Indian official," reported in the Basel Mission Society's *Annual Report*, 1912. United Theological College Archives (UTCA), Bangalore.
2. Eric Stokes's account of Evangelicalism makes a similar argument for the early nineteenth century: "the Evangelical gospel, although originating in an intense interior experience, was one of action and mission in the external world. Work, requiring industry, frugality, and perseverance, was an end in its own right, the outward daily discipline of the soul against sloth; but it also afforded the material means for furthering the Kingdom on earth"; Eric Stokes, *The English Utilitarians and India* (Oxford: Clarendon Press, 1959), 30.
3. For an analysis of imperialism and Utilitarianism in India, see ibid.; for some revisiting and modification of Stokes's claims, see Martin Moir, Douglas Peers, and Lynn Zastoupil, eds., *J. S. Mill's Encounter with India* (Toronto: University of Toronto Press, 1999).
4. Eric Stokes makes a similar point, almost in passing, while discussing the authoritarian and militaristic values within Utilitarianism. He comments, "the key . . . to the emotionalism of imperialism is the transposition of evangelicalism to wholly secular objects, or alternatively the translation of secular objectives to a religious level" (Stokes, *English Utilitarians and India*, 308). In many ways, Stokes's discussion of the commercial and religious aspects of early-nineteenth-century Evangelicalism anticipates my argument here. I attempt to flesh out my argument through the use of material regarding the everyday functioning of selected south Indian missions and the effects on marginalized groups, offering a "ground-level" snapshot of a broader intellectual trend whose early-nineteenth-century intellectual roots Stokes and other scholars have already studied comprehensively.

   My claims in this chapter with respect to the turn of the century are not dependent on Stokes's claim, but the resonances raise interesting research questions about how early to date the interdependence of science and evangelical Christianity. It might also lead us to ask questions about evangelical Hinduism and post-independence science, particularly with respect to India's 1988 nuclear bomb test's significance as a symbol of Hindu nationalist pride.

5. Warren Hastings, 1772, cited in Stokes, *English Utilitarians and India*, 36. Note that Orientalism in this sense refers to the eighteenth-century scholarly discipline that studied "oriental" literature, religion, and philosophy—that is, this is not the sense of the term used by Edward Said.
6. For a comprehensive description of nineteenth-century missions and education, see Gauri Viswanathan, *Masks of Conquest: Literary Study and British Rule in India* (New York: Columbia University Press, 1989).
7. Wood (also known as Lord Halifax) advocated in his dispatch state support for the creation of a system of education that would produce a class of Indians loyal to the British Empire and trained to carry out its clerical and lower-level administrative work. The Wood Dispatch had a significant influence on Indian educational policy, and set the stage for a Western-oriented model that still prevails. For example, the University of Madras traces its origin and mission to the Wood Dispatch.

   Viswanathan quotes from the dispatch on the anticipated effects of the diffusion of European culture and knowledge: "[It] will teach the natives of India the marvellous results of the employment of labour and capital, rouse them to emulate us in the development of the vast resources of their country, . . . and, at the same time, secure to us a large and more certain supply of many articles necessary for our manufactures and extensively consumed by all classes of our population, as well as an almost inexhaustible demand for the product of British labour" (*Masks of Conquest*, 146).
8. See Richard Temple, *India in 1880* (London: John Murray, 1881). Richard Temple's book was written after he had spent thirty years in the Indian administration, during which he had been governor of Bombay, lieutenant-governor of Bengal, and had, he records, been employed under all the departments of the colonial Indian state. Richard Temple's son, R. C. Temple, became a well-known administrator-ethnographer. He founded *Punjab Notes and Queries*, and was editor of *Indian Antiquary*. He was made honorary fellow at Cambridge for his contributions to anthropology, and gave a series of university talks at Birmingham, Oxford, and Cambridge on the practical and administrative value of anthropology.
9. Ibid., 168, emphasis added.
10. Ibid., 174–175.
11. Further, the resulting mixed formations of religious, secular, and economic ideologies characterize a modernity that is still very much with us. This set of formations characterizes modernity exactly because it is not purely a reflection of secular pedagogy, scientific ideals, or capitalist political economy but rather a historically nuanced, sometimes contradictory, mixture of each of these ideological formations with religious thought, romanticist sentiments, and precapitalist modes of production.
12. Address to the Basel Mission Society by "an Indian official," reported in the

1912 *Annual Report* of the society. This unnamed "Indian official" explains that "British perseverance" is exemplified by other socioeconomic spheres and, significantly, he chooses a metaphorical connection between Christianity and the plantation economy to illustrate this point. His comments appear in this chapter's epigraph.

13. By 1900 there were missions in Dharwar, Hubli, Tellicherry, Cannanore, Calicut, Chambala, Gulegud, Belgaum, Coorg, Honnavar, Palghat, Kotagiri, Mercara, Kalkala, Kundapur, Basrur, Bijapur, Kasargod, Puttur, and Kalhatti. Many of the manufacturing units set up by the mission are still in operation today.
14. Basel Mission Society cloth was (and still is) well known in the region. In 1896, a planter's wife, who had a regular column in the journal *Planting Opinion*, wrote: "The cloth from the Calicut Mission makes excellent suits for planters, as they wash well and look well until the very last, and are very cheap . . . and they make them to measure" (*Planting Opinion*, 4 July 1896). When I visited the former Mission Society office in Calicut in 1992, they were making Crate and Barrel table settings and American designer-brand upholstery (under the new name of the incorporated enterprise, The Commonwealth Trust). Weavers sat, sweating, for long days in the huge nineteenth-century warehouse filled with looms, producing for a new global power.
15. See W. E. Albert, "The Basel Mission," in *Basel Mission Souvenir*, edited by Panchikal (Bangalore: Christian Association, 1980).
16. J. Müller, "Industrial Missionary Work," *International Review of Missions* 2 (1913) (UTCA, Bangalore).
17. Basel Mission Society, *Annual Report* 1873 (UTCA, Bangalore).
18. Basel Mission Society, *Annual Report* 1887 (UTCA, Bangalore).
19. *Report of the Basel Mission Society* 1879 (UTCA, Bangalore). It is curious that the writer announces three, and then goes on to name four objects.
20. Anson Rabinbach, *The Human Motor* (New York: Basic Books, 1990), 26–30.
21. It might seem contradictory that there was any way offered out of a racialized, thus supposedly inherent, backwardness. This is a contradiction that characterizes much of nineteenth-century discourse about "others" and also much of current discourse around race, such as that around welfare mothers (they are held to be inherently lazy, ignorant and overfertile, yet are simultaneously the subjects of "improvement" attempts in state social welfare projects). Historians have explained away this contradiction by distinguishing between polygenetic and monogenetic thought within nineteenth-century ethnography; a similar distinction plays itself out today in terms of the "nature-nurture" debate. Lay discourse both then and now has tended to conflate or vacillate between versions of these dichotomies, giving rise to the contradiction as a result of joining discourses with conflicting assumptions. For a useful historical analysis of this contradiction within the discipline, see George Stocking, *Victorian Anthropology* (New York: Free Press, 1987), 238–273.
22. Viswanathan, *Masks of Conquest*, 145.
23. *The Reports of the Protestant Decennial Conferences* are extensively cited in Rev. S.D.L. Alagodi, "The Concept of Mission as Reflected in Protestant Mis-

sionary Conferences in India, 1825–1902," Ph.D. dissertation, Serampore University, 1973.

24. "Work among Indian Outcastes," *1912 International Review of Missions* 1 (1912). Reprinted from an article by the bishop of Madras in *Church Missionary Review* (UTCA, Bangalore).
25. Raymond Williams, *Marxism and Literature* (New York: Oxford University Press, 1977), 86.
26. This and the following quotes are from J. Müller, "Industrial Missionary Work."
27. The history of south Indian Christian mass movements is documented in Alagodi, "The Concept of Mission."
28. Reverend Samuel Mateer, *Native Life in Travancore* (London: London Missionary Society, 1883), 79.
29. Reverend George Matthan, quoted in J. W. Gladstone, *Protestant Christianity and People's Movements in Kerala: A Study of Christian Mass Movements in Relation to Neo-Hindu Socio-Religious Movements in Kerala, 1850–1936* (Trivandrum: KUTS Seminary Publications, 1984), 111 n 173.
30. Reverend Henry Baker, *The Hill Arians of Travancore* (London: Wertheim, Macintosh and Hunt, 1862).
31. See "The Story of the Mission to the Hill Arians," in *Church Missionary Intelligencer*, 1898, 771–773 (UTCA, Bangalore). There was also friction between converted Christian Mala Arayar and those who resisted conversion.
32. Earlier in the century, the raja "had the converts beaten, kept in stocks, doused in water to wash the Christianity out of them, chillies rubbed in their eyes and their heads tied up in bags filled with large black ground-ants and red tree-ants." See Mateer, *Native Life in Travancore*, 79.
33. Reverend Henry Baker, "Pulayas," *Church Missionary Intelligencer*, September 1866 (UTCA, Bangalore).
34. 'A.H.,' *Day Dawn in Travancore: A Brief Account of the Manners and Customs of the People and the Efforts That Are Being Made for Their Improvements* (Kottayam: Seminary Publications, 1860) (Connemara Public Library, Chennai).
35. Ibid.
36. It is commonly noted in popular discourse in Kerala that the Church Missionary Society sacrificed, to a great extent, the interests of the Pulayas in order to keep the Syrian Christians within their fold. The Syrian Christians represented greater financial and racial power (a secular and economic, not a religious or moral concern) to the missionaries, as they were wealthy, educated, and combined an old tradition of Christianity with claims of Brahminical racial origins.
37. Baker, "Pulayas."
38. Quoted in Alagodi, "The Concept of Mission."
39. In a detailed study, Meena Radhakrishna shows how the Yerukulas' criminalization in 1911 provided cheap labor in factories and agricultural settlements, overseen by the Salvation Army. The Salvation Army put down Yerukula workers' mobilization in factories, and opposed Yerukula petitions for the ownership of the land they tilled. Meena Radhakrishna, "Colonial Construction of

a 'Criminal' Tribe: Yerukulas of Madras Presidency," *Economic and Political Weekly* 35.28–29 (2000), 2,553–2,563.

40. Some recent histories of religious nationalism support this argument. See, for example, Tapan Basu, Pradip Datta, Sumit Sarkar, Tanika Sarkar, and Sambuddha Sen, *Khaki Shorts, Saffron Flags* (New Delhi: Orient Longman, 1992). Basu et al. argue that Hindutva is a modern religious phenomenon with roots in the late nineteenth and early twentieth century, which draws on modern notions of the nation-state, and whose specificities have been configured in the mid- and late-twentieth century by capitalist production and modern technological modes of communication and propaganda.
41. Panchikal, ed., *Basel Mission Souvenir* (Bangalore: Christian Association, 1980), emphasis added.
42. Alagodi, "The Concept of Mission."
43. Thompson notes: "Throughout the eighteenth century there is a never-ending chorus of complaint from all the Churches and most employers as to the idleness, profligacy, improvidence and thriftlessness of labour." See Edward P. Thompson, *The Making of the English Working Class* (New York: Vintage Books, 1966), 357–358.
44. Andrew Ure (1835), quoted ibid., 359–360.
45. Rabinbach, *The Human Motor*, 29.
46. Ibid.
47. Ibid., 36, 38.
48. Viswanathan, too, employs a periodization in which religion and morality give way to science and economics in the last third of the nineteenth century. *Masks of Conquest*, 142–146.
49. See, for example, Michael Worboys, "The British Association and Empire," in *The Parliament of Science: The British Association for the Advancement of Science, 1831–1981*, edited by Roy MacLeod and Peter Collins (London: Science Reviews, 1981); Sydney Mintz, *Sweetness and Power: The Place of Sugar in Modern History* (New York: Penguin Books, 1986); Eric Wolf, *Europe and the People without History* (Berkeley and Los Angeles: University of California Press, 1982); Roy MacLeod, "On Visiting the Moving Metropolis," *Historical Records of Australian Science* 5.3 (1982): 1–16.
50. Mateer, *Native Life in Travancore*, 64 .
51. Ibid., 65.
52. Mateer continues, describing the processes by which this race marches on: "The planter approaches them in a peaceable way, offering wages for their hire, but demanding as his right the land he has purchased. The proud men of the woods decline to herd with the coolies, and work like common people. As soon as the planter's axe is heard, the hill kings pack their traps and. . . . In this way they have been driven from hill to hill and from valley to valley, until some have found now a safe resting place in the lowlands of Travancore. If the planter wishes to penetrate some unexplored jungle, or cut a path in some out-of-the-way place, the hill-men are ready to assist, and it is the universal testimony that they are more faithful to their engagements than their more civilized brethren

from the plains" (ibid., 66). This scheme of progress mirrors the one that foresters and planters subscribed to; see chapter 3.

53. See, for example, Frédérique Apffel Marglin and Stephen A. Marglin, eds., *Dominating Knowledge: Development, Culture and Resistance* (Oxford: Clarendon Press, 1990); Ashis Nandy, ed., *Science, Hegemony and Violence: A Requiem for Modernity* (Delhi: Oxford University Press, 1988).
54. David Harvey, *The Condition of Postmodernity: An Enquiry into the Origins of Cultural Change* (Oxford: Basil Blackwell, 1989), 254.

Chapter 7 ***A Global Story***

1. Joseph Reynolds Green, *A History of Botany in the United Kingdom* (London, 1914), quoted in Roy MacLeod, "Scientific Advice for British India: Imperial Perceptions and Administrative Goals, 1898–1923," *Modern Asian Studies* 9.3 (1975): 343–384.
2. Quoted in Susana Hecht and Alexander Cockburn, *The Fate of the Forest* (New York: Harper and Row, 1990), 8.
3. Lucile Brockway, *Science and Colonial Expansion: The Role of the British Royal Botanic Gardens* (London, New York: Academic Press, 1979), 109.
4. Clements Markham, *Travels in Peru and India while Superintending the Collection of Chinchona Plants and Seeds in South America, and Their Introduction into India* (London: John Murray, 1862).
5. The Andean republics exported 2 million pounds of cinchona bark in 1860, 19.84 million pounds in 1881, and 4.41 million pounds in 1884 (Brockway, *Science and Colonial Expansion*), 112.
6. Markham, *Travels in Peru and India*, 66.
7. Ibid., 105.
8. Markham says of him, "M. Hasskarl deserves the greatest credit for the zeal and determination displayed by him in his journeys, during which he was surrounded by no ordinary amount of difficulties and dangers. He certainly proved himself to be a most indefatigable and courageous traveller"; ibid., 51.
9. The sun and the volcanic soil, however, did not prove conducive to the plant, and Markham reports the experiment as "languishing" by 1855. Despite Markham's much feted project, however, the Dutch were ultimately far more successful than the British in manufacturing quinine for the world market. Historians trace this to the fact that Markham ignored an immensely productive species of cinchona, *C. ledgeriana*, named after Charles Ledger, an even more colorful explorer and adventurer than Markham. Ledger, after being turned down by the British, sold his seed to the Dutch in 1865, and by the 1880s Dutch quinine from Java had practically driven British quinine out of the world market; see Richard Harry Drayton, "Imperial Science and a Scientific Empire: Kew Gardens and the Uses of Nature, 1772–1903 (England)," Ph.D. dissertation Yale University, 1993. Brockway, however, points out that market competition was not a priority in the British cinchona project. She argues that "the difference in species is not a sufficient explanation," and that the British and Dutch cooper-

ated in both the scientific and the economic spheres in quinine, as in rubber; *Science and Colonial Expansion*, 120.

10. Markham, *Travels in Peru and India*, 224.
11. Ibid., 276.
12. Joseph D. Hooker was the son of William J. Hooker, first official director of Kew (1841–1855). Joseph Hooker was assistant director of Kew (1855–1865), and served as director from 1865 until 1885, when his son-in-law Thistleton-Dyer took over. See Joseph D. Hooker and Thomas Thomson, *Flora Indica: Being a Systematic Account of the Plants of British India, Together with Observations on the Structure and Affinities of Their Natural Orders and Genera*, vol. 1 (London: W. Pamplin, 1855); and Brockway, *Science and Colonial Expansion*.
13. Hooker and Thomson, *Flora Indica*, 1: 3.
14. Ibid., 13.
15. Ibid., 9.
16. I. H. Burkill's "Chapters on the History of Botany in India," based on his experience, were published in a series in the *Journal of the Bombay Natural History Society* from 1953 through 1960. In 1965 they were published as a book by the Botanical Survey of India, Calcutta.
17. I. H. Burkill, "Chapters on the History of Botany in India," 2–3.
18. Clements Markham, *The Fifty Years' Work of the Royal Geographical Society* (London: John Murray, 1881), 12. The following account of institutionalized geography is from this work.
19. Ibid., 18–20.
20. Ibid.
21. During his tenure, the Rosetta stone and the Elgin marbles were acquired by the Royal Geographical Society for the British Museum.
22. Markham, *The Fifty Years' Work of the Royal Geographical Society*, 37.
23. Brockway, *Science and Colonial Expansion*; Richard Drayton, *Nature's Government: Science, Imperial Britain, and the "Improvement" of the World* (New Haven: Yale University Press, 2000); G. Gramiccia, *The Life of Charles Ledger (1818–1905): Alpacas and Quinine* (London: Macmillan, 1988).
24. The claim that narratives can be mined for their political and social content is not new. The best-known political elucidation of narrative theory is Fredric Jameson, *The Political Unconscious: Narrative as a Socially Symbolic Act* (Ithaca: Cornell University Press, 1981). Jameson suggests not only that it is possible to do political readings of all texts but, further, that "the political interpretation of literary texts" is the "absolute horizon of all reading and all interpretation"; Fredric Jameson, *The Prison House of Language: A Critical Account of Structuralism and Russian Formalism* (Princeton: Princeton University Press, 1972), 17. By this he does not mean that there is no "reality" beyond the text, nor that all ideological readings are equally plausible. He reminds us that "history is *not* a text, not a narrative, master or otherwise, but that, as an absent cause, it is inaccessible to us except in textual form, and that our approach to it and to the Real itself necessarily passes through its prior textualization, its narrativization in the political unconscious" (35).
25. Drayton, "Imperial Science and a Scientific Empire," 344.
26. Brockway, *Science and Colonial Expansion*, 113.

27. This was discussed in chapter 2. For a historical account of these attitudes to nature, see Clarence J. Glacken, *Traces on the Rhodian Shore: Nature and Culture in Western Thought from Ancient Times to the End of the Eighteenth Century* (Berkeley and Los Angeles: University of California Press, 1967).
28. Dane Kennedy, *The Magic Mountains: Hill Stations and the British Raj* (New Delhi: Oxford University Press, 1996); Kavita Philip, "English Mud: Towards a Critical Cultural Studies of Colonial Science," *Cultural Studies* 12.3 (1998), 300–331. The hill station was a place in the hills where troops and administrations could be "stationed," most commonly in the hot summer months when the weather in the plains became unbearable for the British.
29. For an example of discourse analysis from literary studies, see Mary Louise Pratt, *Imperial Eyes: Travel Writing and Transculturation* (New York: Routledge, 1992); for examples of studies of the politics of environmentalism, see William Cronon, *Nature's Metropolis: Chicago and the Great West* (New York: Norton, 1991), and Brockway, *Science and Colonial Expansion*.
30. Markham, *The Fifty Years' Work of the Royal Geographical Society*, 5.
31. Charles de La Condamine and other European botanists were not, however, saying that non-Europeans could never contribute to colonial science. Markham gives several examples of native assistants who produced botanical knowledge, once having been introduced to the professional methods of observation and recording.
32. Hecht and Cockburn, *Fate of the Forest*, 8.
33. See Carolyn Merchant, *The Death of Nature: Women, Ecology, and the Scientific Revolution* (San Francisco: Harper and Row, 1990), 168–170; in her reading of Francis Bacon's work, she characterizes scientists as putting nature on a torture rack to extract her secrets. It is interesting that Markham represents the natives as doing something similarly violent to nature; colonial scientists are represented as nature's saviors.
34. Markham, *Travels in Peru and India*, 25.
35. Ibid., 44.
36. Ibid., 46.
37. Quoted in Hecht and Cockburn, *Fate of the Forest*, 9.
38. Markham describes the French botanist Weddell as an "able botanist and intrepid explorer [to whom] science is indebted to no small extent." His first "cinchona trip" to Peru was in 1845, and his 1849 *Histoire naturelle des quinquinas* was "the most important work that has yet appeared on the subject," according to Markham (see *Travels in Peru and India*). The French were attempting to cultivate cinchona at the Jardin des Plantes in Paris. We have already noted the expedition of the Dutch botanist Hasskarl.
39. Markham, *Travels in Peru and India*, 60.
40. Ibid.
41. Cited in Brockway, *Science and Colonial Expansion*, 103.
42. Captain Campbell Walker, deputy conservator of forests, *Report on Government Cinchona Plantations on the Nilgiris*, 1878. G.O. Revenue Department No. 72, 15 January 1878 (TNSA).
43. As Brockway points out, however, the coppicing technique adopted in the Nilgiri

plantations was "consistent with the native Andean harvesting practice" (*Science and Colonial Expansion*, 119).

44. See Richard Grove, *Green Imperialism: Colonial Expansion, Tropical Island Edens and the Origins of Environmentalism, 1600–1860* (New York and Cambridge: Cambridge University Press, 1995).
45. Markham, *Travels in Peru and India*, 507.
46. Man was believed to be at the center of nonscientific systems of knowledge, while the position of the observer was irrelevant in scientific systems. See Burkill's discussion, earlier in this chapter. See also Michel Foucault, *The Order of Things: An Archæology of the Human Sciences* (New York: Vintage Books, 1973), especially pages 144–160. Foucault points out that the language of taxonomy depended on an objective classificatory system, and was distinguished from nonscientific language, which "leave[s] insterstices open" between scientific propositions and designations, which might then be inhabited by "individual experiences, needs or passions, habits, prejudices" (158). If the play of language was restricted, one could see without error.
47. Tarun Chhabra, "Cinchona Forests: Sinking Rapidly," *The Hindu*, 6 June 1993.
48. Wolf, *Europe and the People without History*.
49. Jack Kloppenberg, *First the Seed: The Political Economy of Plant Biotechnology* (Cambridge: Cambridge University Press, 1988), 15.

Chapter 8 ***Conclusion***

1. Karl Marx, *Economic and Philosophic Manuscripts of 1844*, cited in Robert C. Tucker, ed., *The Marx-Engels Reader*, 2d ed. (New York: W. W. Norton, 1978), 90.
2. See Johannes Fabian, *Time and the Other: How Anthropology Makes Its Object* (New York: Columbia University Press, 1983).
3. Recall Markham's rhetoric of universal benefit, and see John Berger, *About Looking* (New York: Pantheon Books, 1980), chapter 1, for a discussion of the connection between marginalizing nature and romanticizing it.
4. Recall that in chapter 2, I argued that the poem "Nilgiri Sunshine" could be read as eulogizing a pure, elevated nature along with its pair, a sophisticated culture, implicitly contrasting this pair with its mirror-opposite pair, namely, low or debased nature and low or primitive culture. In chapter 4, I discussed how amateur ethnographer W. E. Marshall's description of the Todas as oblivious to the beauties of their surroundings and simultaneously incapable of labor demonstrates a similar dichotomy.
5. T. J. Barron, "Science and the Nineteenth-Century Ceylon Coffee Planters," *Journal of Imperial and Commonwealth History* 16.1 (1987): 5.
6. Gauri Viswanathan, *Masks of Conquest: Literary Study and British Rule in India* (New York: Columbia University Press, 1989), 127.
7. Michael Adas, *Machines as the Measure of Men: Science, Technology, and Ideologies of Western Dominance* (Ithaca: Cornell University Press, 1989), 340.
8. As epigraphs, of course, they are out of their contexts; within their own definitions of race, morality, literature, or science, each of them is indeed self-

consistent. However, these definitions are, in all three cases, narrowly construed and quite inadequate to deal with the scope of the actual workings of these discourses in their full complexity, to say nothing about their mutual intersections. Barron sees science as a body of theoretical knowledge existing outside the field of political and economic power; Viswanathan uses a dichotomous model that identifies science with secular and practical activity, and excludes moral and philosophical theorizing; Adas sees racism as contained solely in discriminatory policies, and excludes from it such practices as the construction of essentialized identities. Barron and Viswanathan thus have limited definitions of science, while Adas has a limited definition of the social meanings of race. I have addressed Barron's and Viswanathan's arguments with some specific examples in chapters 3 and 5, and Adas's rather simplified model of race can easily be problematized with reference to a number of recent studies of race and society. See, for instance, Sandra Harding, *The "Racial" Economy of Science: Toward a Democratic Future* (Bloomington: Indiana University Press, 1993); Henry Louis Gates, ed., *"Race," Writing and Difference* (Chicago: University of Chicago Press, 1986); Dominick LaCapra, ed., *The Bounds of Race* (Ithaca: Cornell University Press, 1991); David Wellman, *Portraits of White Racism* (New York: Cambridge University Press, 1977). I will not, therefore, dwell at length on the specifics of their claims, as my aim is not to deny the very useful contributions they (Adas and Viswanathan in particular) have made to the historical study of colonialism in South Asia.

9. To illustrate with an analogy: in critiquing the exploitative and patriarchal aspects of Enlightenment thought, feminist and postcolonial critics have not ceased to live in a world saturated with Enlightenment-derived practices, institutions, and ways of thinking. For the coinage of the term "situated knowledges," see Donna J. Haraway, "Situated Knowledges: The Science Question in Feminism and the Privilege of Partial Perspective" in *Simians, Cyborgs, and Women: The Reinvention of Nature* (New York: Routledge, 1991), 183–201.
10. Sandra Harding, *Whose Science? Whose Knowledge? Thinking from Women's Lives* (Ithaca: Cornell University Press, 1991), 211.
11. Ibid., 151.
12. For an assessment of Mertonian, Polanyian, and strong constructivist models of sociology of science with respect to their capacity to make critical interventions in science and society, see Joseph Rouse, "What Are Cultural Studies of Scientific Knowledge?" *Configurations* 1.1 (1992): 1–22.
13. It pays to remind ourselves at this point that assertions regarding the mutual construction of science and society do not entail the position that social processes and political motives completely determine the content of scientific theories. Knowledge is, while irreducibly social, always underdetermined by the social forces that go into its production. The residue is something that realist philosophers have reminded us not to rule out, namely reality. As Sismondo says, realism "serves as little more than a reminder not to rule out the material world and its structures when we are investigating the processes and powers of science. . . . [T]ruth-talk has at least potentially some political value"; Sergio Sismondo, "Some Social Constructions," *Social Studies of Science* 23.3 (1993), 361.

Index

Page numbers in italics indicate illustrations

Académie des Sciences, 185
Adas, Michael, 235n8
administrative control, British, 25, 111, 113, 135, 151–152
Administrative Report on Forests in Madras Presidency, 63, 64, 74, 212n21, 214n56
Africa, wildlife in, 17–18
African Association, 177, 178
Agency Tracts, 21
Agricultural and Horticultural Society of Calcutta, 97
agricultural experiment stations, 94
Alagodi, Reverend, 163
Alderson, E.A.H., 18
anachronism, concept of, 13
Andaman Islanders, *106*
Andean region, 171, 172–173, 191, 231n5
Another Reason (Prakash), 27
Anthropological Institute, 101
anthropology, 107–109, 123–124, 196, 218n6; amateurs in, 103, 125; biblical, 220n35; colonial, 100–104, 134, 135, 197; and criminality, 127–128; critique of, 217n3; indigenous practitioners of, 139–143; physical, 138; status of, 121–122; and time, 219n23. *See also* ethnography
anthropometry, 105–107, 126–128, 140–141, 165, 222n67. *See also* phrenology
Army, Indian, 128
ascetics, 131
Asiatic Society of Bengal, 108
Australia, 38
authority, of colonizing groups, 33, 38–40

Badagas (tribe), 36, *37,* 38, 41, 44, 71–72, 99, 226n111
Baden-Powell, B. H., 61–62
Baker, Henry, 159, 160
Banks, Sir Joseph, 177
Barron, T. J., 97–98, 235n8
Basel Mission Industrial Committee, 158, 162, 228n14
Basel Mission Society, 69, 144, 152–155, 157, 158, 162–163, 163
Basel Mission Souvenir, 163
Basel Mission Trading Company, 153
BBC talk show, 91
beauty, appreciation of, 116–117, 122
beggars/mendicants, 131, 223n72
Belgium, 85
Bengal, 25–26, 27, 173
Bennett, Tony, 136, 224n86
Bible, 155
biblical anthropology, 220n35
biblical ethnology, 114, 116–117, 156

Bidie, Surgeon-Major, 188
birth rate, among natives, 41
Blair, G. W., 115
Board of Agriculture, 97
Bolivia, 174, 186, 190
Bombay Natural History Society, 20–21
bonded/debt labor, 65, 70, 81
botanical gardens, 89, 97, 181
Botanical Survey, 177
botany, 6, 11; global networks of, 53, 171–172, 194–195; imperial, 181; institutional development of, 177–179; systematization of, 176–177, 233n31. *See also* cinchona tree
Boyd, Richard, 12
Brahmins, 13, 105, 106, 107, 114. *See also* elite, indigenous; upper-caste groups
Brandis, Dietrich, 211n12, 212n14
British Association for the Advancement of Science, 101
British East India Company, 26, 59–60, 149, 182
British Parliament, 26
British Raj, 26–27, 54, 127, 131, 149, 206n38
Brockway, Lucile, 181, 231–232n9, 233–234n43
Brown, Charles Hilton, 209n12
buffalo, 41, 44
bureaucratic restrictions, British, 40. *See also* administrative control
Burkhill, I. H., 177
Burton, Reginald, 17

Calcutta School of Tropical Medicine, 94
Calicut, 24–25
capitalism, history of, 1
Caravaya forests, 174–175
Carson, Penelope, 26
cascarillas (bark harvesters), 189–190
cascarilleros (bark collectors), 175, 190. *See also* cinchona tree
caste divisions, 13, 70–71, 108, 152, 156–157. *See also* class stratification; cultural differences/positions
caste "personalities," 128
Catholic missions, 11, 83
cattle, 57, 71–72
Chanda, Ramaprasad, 140–142
Chapparbands (tribe), 132
Charter Act of 1813, 149
Chenchus (tribe), 55, 57, 64–66, 75, 99
Chetties (tribe), 25, 70, 83, 208n7
children, and separation from parents, 130
Christianity, 115, 147–170; and Malabar tribes, 159–162; and modernity, 162–164, 166–170; narratives of, 118, 119–121; resistance to, 44; science's influence on, 148–152; and work, 152–159. *See also* missionaries; missions; religion, and science
Christian Missionary Society, 159–162
Church Missionary Society, 229n36
Cinchona Committee, 192
cinchona tree, 53, 171–175, *173,* 180–193, 231–232n9; bark of, 175, 185, 186, 189–190, 231n5; economics of, 187–189; and environmental narratives, 180–182, 193–195; history of, 171–175; *ledgeriana* species, 180–181, 231–232n9; moral rhetoric around, 189–191; and systems of knowledge, 183–186, 191–193
civilization, progress/status of, 64–72, 105, 117, 121, 122–124, 148–151, 155, 166, 183; and caste stratification, 70–71; hierarchy in, 97, 137; resistance to, 64–65; and resource use, 66–67
civil servants, 136. *See also* administrative control
class stratification, 150. *See also* caste divisions
clear-felling, 67
climate, Nilgiri, 33–35
Clive, Robert, 25–26
coal walking, 44
Cockburn, Alexander, 185
coffee cultivation, 39, 52, 120

Coimbatore, 31, 60
Cole, Inspector-General, 50–51
collahuayas (traveling healers), 175
colonial discourse analysis, 169–170
colonial environmental history, 180
colonial history, of India, 24–28
colonialism, assumed function of, 138–139
colonial modernity, 12–13, 78–79. *See also* modernity
colonial power, and science, 97–98, 102–104, 118, 120–121
colonial science, 4–8, 10, 11, 14; critical studies of, 42, 198–201. *See also* science, and colonialism
colonial subjects, 4, 7–8. *See also* indigenous people; natives; tribals
colonies, as laboratories, 6
commercial enterprises, 23. *See also* forestry, in Madras; manufacturing; plantations
Condamine, Charles Marie de, 184–185, 233n31
Conditions of the Working Class in England, The (Engels), 138
conservation, 20–21, 59, 60, 62, 180, 190, 206n35. *See also* natural resource management/use
coolies (plantation workers), 80
Cooly Mission, 69
"cooly missions," 120
Coombes, Josiah Waters, 127
coppicing technique, 233–234n43
Corumber, Curumber (tribe). *See* Kurumbar
Cox, S., 69
criminalization, of tribes, 22, 126–134, 143–144, 162. *See also* resistance, native
Criminal Tribes Act (1871), 22, 64, 128, 129–130, 133, 134, 222–223n69
Criminal Tribes Act in Madras Presidency (1911), 128
cross-context knowledge, 192
Crown rule. *See* British Raj
cultivation, narratives of, 50–53, 67, 115, 166. *See also* slash-and-burn cultivation
cultural differences/positions, 134–135, 137, 197–198, 234n4. *See also* caste divisions; class stratification
cultural diffusion, lag in, 88
cultural materialist historiography, 9–14
cultural studies, 9–12, 42
culture, science as, 180
culture/nature dichotomy, 20, 36, 117, 197–198, 234n4

Daniel, P. H., 81–82, 94, 215n6
Darwin, Charles, 57, 211n9
Das Gupta, Ranajit, 81
debt/bonded labor, 65, 70, 81
Decennial Conference of Protestant Missions (1879), 161
deforestation. *See* forestry
deities, tribal, 38, 44, 45, 220n43
"denial of coevalness," 219n23
development models, of India, 7, 25
disease, 42, 62–63, 71; malaria, 50–51, 53, 80
District Planter's Associations of South India, 93
Donga Yerakalas (tribe), 129
doraies (bosses), 80
Dravidian race, 141
Drayton, Richard, 181
drunkenness/drinking, 57, 75, 85–86, 118
Durbar Day, 77
Dutch East India Company, 25
Dutch Empire, 173, 180–181, 194, 231n9

East India Distilleries and Sugar Factories, Ltd., 134
ecofeminism, 205n26
economic development, of forests, 78
economic progress, defined, 190
economic rationality, 163, 187
economic systems, 59, 170, 186, 189, 190, 192. *See also* globalization/global trade
education, 26–27, 67–69, 122–125, 150, 151, 158. *See also* schools

Education Dispatch of 1854, 150
elephants, 48, 71–72
elite, indigenous, 26, 27. *See also* Brahmins; upper-caste groups
Elphinstone, Mountstuart, 36, 178
Elwin, Verrier, 8
Emeneau, M. B., 145, 226n113
Engels, Friedrich, 137–138
engineering, and forestry, 76–77
Enlightenment models of science, 8
environmental history, 4
environmentalism, 6
environmentalist narratives, 63, 72–73, 179–182, 193–195. *See also* nature
Ernâdens (tribe), 105
ethnography, 21, 99–146; and anthropology as science, 100–104, 108–111; and criminal tribes, 126–134; and ideologies of modernity, 23–24; and indigenous anthropologists, 139–143; and labor, 70–71, 83–84; and missionaries, 114–122; moral and political significance of, 111, 113, 122–126, 134–139; and native resistance, 143–146; and race, 104–108, 162
Ethnological Survey of India, 113
Europe and the People without History (Wolf), 12
Evangelical movement, 149, 226nn2, 4
exploitation, vs. romanticism, 6–7
exports, Indian, 96

Fabian, Johannes, 217n3, 219n23
factories, 133–134, 164. *See also* manufacturing
famine, 18, 68, 125–126, 160, 161, 222n61
Fawcett, Fred, 105–106
febrifuge (quinine source), 189
feminist scholarship, 199
First War of Independence, 26
Flora Indica (Hooker and Thomson), 176
folk songs, 144
Forest Administration Reports, 73
forest clearing, 51
Forest Department, 47, 54, 61, 120; changes in, 76–77; and labor, 63–73; preparatory schools, 67–68; and tribal resistance, 74. *See also* forestry, in Madras
Forest Department Proceedings, 72, 73
forest management programs, 39
forest officers, 48–50, 159–160
forestry, in Madras, 21, 23–24, 54–79; administration of, 55–57, 214n56; and changes in landscape, 221n49; disciplining labor in, 63–73; global issues in, 78–79; local vs. colonial use, 55–61, 66–67, 102; and native education, 64, 66, 67–69; native resistance and, 54, 64–65, 73–78; and nature as property, 61–63; timber clearing/felling, 67, 74
forests, changes in, 221n49
forest village, 59, 60–61
France, 194; trading with India, 25
Frohnmeyer, L. J., 153
funeral songs, 43, 144

Gall, Franz Joseph, 138
Gama, Vasco da, 24
Gandhi, Mohandas K., 75
gardening, 32–33, 209n9. *See also* botanical gardens
Gayer, C. W., 130–132
Geddes, Patrick, 8
geographical knowledge, 182
geography, 178
George III, king of England, 26
Ghani, Muhammed Abdul, 128–129, 133
global exploration, in botany, 53, 176–177
globalization/global trade, 59, 95–96, 186, 190, 203n2; in forestry, 24–25; history of, 1, 11; and labor practices, 166; and modernity, 78–79. *See also* cinchona tree
global knowledge, vs. local knowledge, 168, 191–192
Goldsbury, C. P., 91
Gover, Charles, 225–226n111

governmental power, 62. *See also* administrative control
Government of India Resolution (1894), 75–76
Grove, Richard, 21, 63, 180
Gundert, Hermann, 153

Hakluyt, Richard, 182
Hall, G. Stanley, 104
Hamilton, William, 178–179
Harding, Sandra, 199–200
Harkness, Henry, 109
Haruvaru (tribe), 44
Harvey, David, 168
Hasskarl, Justus Charles, 174, 231n8
Headrick, Daniel, 87–89, 139
healing, indigenous, 175, 184
Hecht, Susana, 185
hegemonic narratives, of nature, 31, 208n5
Henry, E. R., 128
Hill Arians, 159–161. *See also* Mala Arayar
hill station, 33, 39–40, 233n28
hill tribes, 39, 40–41, 44, 63–64, 230–231n52. *See also specific tribes*
Hill Tribes of Travancore, The (Matthan), 159
Hinduism, 115, 137, 227n4
Hindu Missionary Societies, 131
Hindus, upper-caste, 141, 152
Hindutva, 230n40
History of British India (Mill), 135, 136–137
"homeless races," 58. *See also* nomadic populations
Homo Alpinus, 141
Hood, H. M., 65–66
Hooker, Joseph D., 176, 181, 232n12
How to Evolve a White Race (Iyer), 142–143
Human Motor, The (Rabinbach), 84, 215–216n11
human progression, theories of, 58
Humboldt, Alexander von, 172, 184
Hunt, James, 138
Hunter, Alexander, 122–125
Hunter, Sir William, 115
hunting, 17–18, *18, 19,* 20–21, 46–47
Huxley, T. H., 101
hygiene, and labor, 85–86

idleness, narratives of, 155–156, 165, 215–216n11, 216n15. *See also* laziness, narratives of
idolatry, among natives, 45, 46
immigration, from plains, 64
Imperial Council of Agricultural Research in India, 96
Imperial Departments of Agriculture, 181
Imperial Institute, 94
imperialism, "new" vs. "old," 87–89
India Act XIII (1859), 92, 93
India Forest Act (1878), 60
Indian Forester reports, 57, 58, 62, 70
Indian Forest Service, 60
Indian National Congress, 27
indigenes, as term, 204–205n14
indigenous knowledge, 6, 16–17, 20, 183–186, 191–193. *See also* knowledge; local knowledge
indigenous narratives, 36, 45
indigenous people, 15–16, 204–205n14. *See also* colonial subjects; natives; tribals
indigenous resistance, 38. *See also* resistance, native
Industrial Revolution, 3–4, 164
industry, and missions, 158–159. *See also* factories; manufacturing
International Congress of Tropical Agriculture, 85, 94
interpretive history, 13
ippa tree, 57, 67
Irulas (tribe), 14
Iyer, L. A. Krishna, 140
Iyer, L. K. Ananthakrishna, 107, 140, 142
Iyer, S. Sundaresa, 142–143, 225n105

Jameson, Fredric, 232n24

Java, 173
"Jesuit's powder," 172
Jews, 105
Juvenile Criminal in Southern India, The (Coombes), 126–127

Kadirs (tribe), 15, *56,* 142
Kanara, 25
Kanikars (tribe), 118, *119,* 120, 142, 166–167, 221n46. *See also* hill tribes; Mala Arayar
Karve, Travati, 222n67
Kerala, 140
Kew Gardens, 11, 53, 173, 177, 181, 183, *184,* 188
Khonds (tribe), 74–75
King, Anthony, 33
Kistnamachari, V., 124, 125
Kloppenburg, Jack, 194
knowledge: critical studies of, 198–201; indigenous/local, 6, 11, 16–17, 20, 113, 195; production of, 104, 123–126, 163–164; scientific, 122, 134–139, 161–164, 172, 234n46; systems of, 166–170, 182, 183–186, 191–193
Koravas (tribe), 133, 134, 143–144
Koyas (tribe), 56, 64, 67, 75
Kuklick, Henrika, 101–104, 219n27
kumri, 49, 50. *See also* slash-and-burn cultivation
Kurumbar/Corumber (tribe), 48, 51, 69, 82, 99, 142; and ethnography, 105–106, 121; narratives of, 47, 83–84; subtribes of, 70–71

labor, 51–52, 59, 216n15; and begging, 223n72; as civilizing, 116, 117, 122; coercive conditions in, 81–82, 97; colonial discourses on, 84–87, 162–163; and education, 67–69; ethnographic information on, 83–84; feudal, 65, 86; Madras tribes as, 63–73; morality of, 84–85, 131, 154–156; native resistance to, 64–65; patterns in, 70–71; plantation, 23, 81–87, 91–94; practices, 164–166; and proprietary interest, 38–40; regulations about, 93–94; slavery, 83, 159; typology of, 47–48; wage, 22–23, 55, 65. *See also* work
laboring class, 155, 156, 163, 215–216n11
La Capra, Dominick, 13
landscape narratives, 30–31, 45–46. *See also* nature: constructions/narratives of
land tenure systems, 38–39
land use, 21. *See also* local use, of resources; natural resource management/use
La Peyrere, Isaac de, 114
laziness, narratives of, 49, 52, 110, 115–116, 165, 166. *See also* idleness, narratives of
Leghorn, Hugh, 60, 212n14
Ling, Catherine, 45
Linlithgow Commission Report, 96
Linnaeus, Carolus, 173
local knowledge, 6, 16–17, 168, 191–193, 195. *See also* indigenous knowledge; knowledge
local use, of resources, 21, 55–61. *See also* forestry, in Madras: local vs. colonial use; natural resource management/use
London Daily Mail, 90–91
London Missionary Society, 115, 118, 133, 166
lower-caste groups, 156–157, 160, 162–163
lying, to colonial officials, 42–44
Lytton, Lord, 29, 30

Macaulay, Thomas Babbington, 26, 29–30, 150, 207–208n3
Machiwanyika, Jason, 18
MacKenzie, John, 18
Madras, 29, 30, 77, 189. *See also* forestry, in Madras
Madras Act V (1866), 92, 93
Madras Forest Act (1882), 60–61, 211–212nn12–13
Madras Government Museum Bulletin, 110

Madras Planters Act (1903), 94, 215n6
Mahabharata, 141–142
maistries (plantation foremen), 80, 81
Mala Arayar, 159–160. *See also* hill tribes; Kanikars
Malabar, 47, 60, 73–74, 81, 105–106; tribes, 159–162
Malabar traders, 25
malaria, 50–51, 53, 80
Malayalees/Malialis (tribe), 74, 214n50
manufacturing, 153–154, 228n13. *See also* factories
Manusmriti, 107
Mapilla rebellion, 73–74
market economy/forces, 59, 186, 189, 190
Markham, Clements, 173–176, 182–188, 190, 191, 192, 231nn8–9, 233nn31, 33, 38
Marshall, William E., 109–110, 114, 116, 122
Martel, Don Manuel, 174–175
Martin, C. S., 76, 78
Martinez, Mariano, 175
Marx, Karl, 137, 138–139
Marxist historiography, Indian, 27
mass conversion movements, 159, 160
Mateer, Samuel, 115–116, 118, 120, 166–167, 230–231n52
Matthan, George, 159
medical missions, 161, 163
Melchett, Lord, 95–96
Merchant, Carolyn, 233n33
Methodism, 164
Metz, Friedrich, 41, 42–44, 45, 46, 114, 144, 225–226n111
middle-caste groups, 156
middle class, creation of, 150
Mill, James, 135, 136–138
Minchin, A. F., 69
mineral products, 221n54
miners, tribals as, 133
mining, 212n21
Minute on Education (1835), 150
missionaries, 23, 42, 118, 120–121, 148–152; and criminal tribes, 133; and ethnography, 114–122; financing of, 17
missions, 11, 68–69, 228n13; education in, 143–144, 158; and ideologies of modernity, 23–24; industry in, 153–154, 158–159; and labor, 154–156; state support of, 151. *See also* Christianity; religion, and science
modernity, 25, 72–73; constructions of, 7, 166–170, 227n11; critiques of, 15–16, 17; effects of, 27–28; and globalization, 78–79; ideologies of, 23–24; mixed models of, 87, 88–89, 97–98, 162–164; native appropriations of, 14–16; and power, 118, 120–121; and technology, 152–155, 157–159
Mohanty, Satya, 12, 136
monocultures, 23
monopoly, timber, 60
morality, narratives of, 97–98, 127, 162, 163, 166, 189–191; and labor, 84–86, 87, 131, 154–156, 158–159, 215n11, 226n2; and progress, 62–63, 68–69, 124. *See also* Christianity; missionaries; missions
Morgan, H. R., 47
morphological characteristics, 165. *See also* anthropometry; phrenology
Morrison, Charles, 103
Mukurti Peak, 116, 220n43
Mullaly, Frederick, 127–128
Müller, J., 154, 158
multidisciplinary studies, 2–3
Munro, Sir Thomas, 60
museums, 224n85
Muslims, 25
"Mutiny" of 1857, 26, 206n38

Nairs Christians, 160, 161
Nambûtiris (tribe), 105
narrative theory, 179–180, 232n24
nationalism, Indian, 27, 139, 140
Native Life in Travancore (Mateer), 115, 118, 166
natives: appropriations of modernity by, 14–16; attitudes toward, 41–42, 100,

natives (*continued*)
123, 206n38; constructions/narratives of, 20, 23, 36, 197–198; as labor, 47–50, 91–94; resistance of, 22–23, 42–44, 54, 55–56, 64–65, 143–146; threat of, 20, 206n34; use of term, 204–205n14. *See also* indigenous people; tribals
natural history, 21, 23; and ideologies of modernity, 23–24
natural resource management/use, 21, 100–102, 169, 172, 185–188, 194. *See also* conservation; local use, of resources
nature: capacity to appreciate, 116–117; in colonial history, 17–23; constructions/narratives of, 23, 30–31, 35–36, 40–41, 46–53, 55, 182, 208n5; and cultural materialism, 9–12; and culture, 20, 36, 117, 197–198, 234n4; and planters, 89–92, 216n25; as property, 61–63; scientist as defender of, 185–188, 190–191, 233n33; use of term, 204–205n14
New York Tribune, 138
Nilgiri Hills, 29–53; and cinchona tree, 171, 172, 173, 181; climate of, 33–35; compared to England, 29–31; early English settlement in, 31–33, 36–40; and ethnography, 105–106, 108–109; global significance of, 52–53; health, labor, and planting in, 50–52; natives of, 39, 41–44, 116; nature in, 35–36, 44–50; population of, 81
"Nilgiri Sunshine" (poem), 33–36, 40, 220–221n44, 234n4
"Nisada" (race), 141–142
nomadic populations, 22–23, 129–131, 166. *See also* "homeless races"
non-Western knowledge, 11–12. *See also* indigenous knowledge; local knowledge
noses/nasal indices, 110, 111, *112*
Note on the Settlement of Criminal Tribes (1925), 143–144
Notes on Criminal Classes of the Madras Presidency (Mullaly), 127–128
Notes on Criminal Tribes of the Madras Presidency (Ghani), 128–129

objectivity, 135–136, 176–177, 200, 234n46. *See also* scientific knowledge: observer and
Olivier, Sir Sydney, 85–86
Ootacamund (Ooty), 29–30, 41, 47, 50, 99, 208n4; early English settlement in, 31–33, 36–38
Orientalism, 149, 150, 227n5
Osorio, Ana de, 172
"outcastes," 159. *See also* tribals

Pálaul (tribe), 41
"Pamirian race," 141
Paniyans (tribe), 14, 70
Parry and Company, 134
paternalism, feudal, 86
paternalistic narratives, 49–59, 66, 84, 91
Patharries (tribe), 132
Pearson, M. N., 25
pedagogical efforts, toward natives, 41–42, 155. *See also* education; schools
Peradeniya Botanical Gardens, 89
Peru, 173, 174, 191
petitions, protest, 75–76
Philosophy of Manufactures (Ure), 164
Phrenologist amongst the Todas, A (Marshall), 220n35
phrenology, 109, 114, 122, 138
Pitt, William, 26
plantations, 69; economy of, 87–89; and ideologies of modernity, 23–24; labor in, 81–87, 92–94; as metaphor, 147; and model of modernity, 97–98; and planters' status, 90–91; private, 50, 51, 181–182, 212n21; and relationship to nature, 89–90, 216n25; science and global trade in, 94–97
planters' associations, 83, 90, 93, 94, 134, 207n40
Planters' Chronicle, 85–86, 92, 94–95
Planting Opinion, 82

Playne, Somerset, 83
police, 127–128, 131
political administration, and ethnography, 111, 113, 134–139. *See also* social policy, and ethnography
political economy, 150, 170
political power, 5–6, 62
political significance of work, 156–159
Porteous, Colonel (inspector-general of police), 128
Portuguese traders, 25
postcolonial development, of India, 7
postcolonial historiography, 14–17
postcolonial studies, 199
"post-positivist realism," 12
poststructuralist narratives, 205n24
Povinelli, Elizabeth, 38
Prakash, Gyan, 27
premodern elements, retention of, 72–73, 78–79, 81, 97. *See also* traditional cultures, preservation of
private contractors/entrepreneurs, 50, 51, 60, 181–182, 212n21
private property, 39, 50, 58, 61–62, 168–169. *See also* land tenure systems
Proceedings of the Chief Conservator of Forests, 55
production practices, 58, 157–159, 168, 169
progress, narratives of, 13, 22, 29, 58, 97; and civilization, 78, 116, 122–125, 148–151, 150, 151, 155, 158; economic, 163, 190; moral, 68–69, 124; scientific, 42, 52, 183. *See also* modernity
propaganda, supporting Forest Department, 73, 77
Protestant mission activity, 11, 69, 163. *See also* missions
proverbs, tribal, 144
publishing companies, 153
Pulaya schools, 160, 161
Punniars (tribe), 48

quinine, 53, 64, 173, 174, 183, 184, 188, 189, 231n9. *See also* cinchona tree

Rabinach, Anson, 84, 155, 164–166, 215–216n11, 216n15
race, classifications/concepts of, 86, 104–105, 106, 107, 142–143, 228n21, 235n8. *See also* anthropometry
racial barriers, 90. *See also* caste divisions
"racialized" discourse, 136, 137–139, 156, 198–199
racial origins, Indian, 140–142
Radhakrishna, Meena, 229n39
railways, 26
Raj, British, 26–27, 54, 127, 131, 149, 206n38
raja of Travancore, 160, 229n32
Raleigh Club, 177–178
Ramayana, The (Valmiki), 131
Rangachari, K., 110, 119, 219n27
recreational activities, 90
Reddis (tribe), 75, 129
Red Tea (Daniel), 82, 215n6
reformatories, 127, 222n62
religion, and science, 19, 97–98, 148–152, 162–164, 170, 227n4
religious instruction, 127. *See* missions: education in
religious mendicants, 131
religious texts, 108
Report of the Cinchona Committee (1878), 189
resistance, native, 22–23, 38, 42–44, 54, 55–56, 64–65, 73–78, 143–146
resources. *See* natural resource management/use
Revenue Department, 61, 63, 65, 76
revenue systems, land, 26
Rise of Anthropology in India, The (Vidyarthi), 121
Risley, Herbert, 113, 128, 130, 140, 141, 222n67
Rivers, William H. R., 110, 140, 208n6, 219n26
roads, public, 52
romanticist narratives, 6, 8, 15–16, 45–47
Ross, Sir Ronald, 85, 216n16
Rouse, Joseph, 200

Roy, Francisco, 31
Roy, Sarat Chandra, 140
Royal Geographical Society, 177–178, 182
Ruiz, Hippolito, 184, 185

Sadhus, 131
Salvation Army, 133, 161, 162, 229n39
Savaras (tribe), 77–78
School of Industrial Arts, 123, 124
schools, 64, 66, 67–69, 143–144, 213. *See also* education
science: amateur organizations in, 178, 179; and arguments for colonial power, 97–98, 102–104, 118, 120–121; as defender of nature, 185–188, 190–191, 233n33; Indian nationalists in, 139–143; models of, 235n8; in plantation economy, 94–98; and religion, 148–152, 162–164, 170, 227n4; and social constructions, 196–197, 235n13; studies, 28, 180, 201; terms in, 207n42; tribal, 132–133
science, and colonialism: concurrent development of, 3–8, 134; cultural materialist studies of, 9–12; methodology in, 12–14; political context of, 21–24; and postcolonial historiography, 14–17
Science and Culture (journal), 140
scientific discourse, of progress, 183
scientific exhibitions, 136, 224nn85–86
scientific knowledge, 122, 172; effects of, 134–139; and missionary work, 161–164; observer in, 191–192, 234n46; vs. local knowledge, 166–168, 183–186, 191–193. *See also* objectivity
scientific modernity, 7, 12–13, 16–17, 25, 72–73, 78–79, 97, 118, 120–121, 162–164. *See also* modernity
Scudder, Ida, 161
Seattle, Chief, 194
sedentarization, forced, 22–23, 64, 129–130, 133–134, 162, 166
sexual harassment, 215n7
shikar, 19, 21, 46. *See also* hunting
Sholagas (tribe), 68, 71–72
Shona uprising, 18
silviculture, 67, 76, 78. *See also* forestry
sin, native forms of, 43
Siraj-ud-Daula, 25–26
Sismondo, Sergio, 235n13
skin color, 142–143. *See also* race, classifications/concepts of; racial barriers
slash-and-burn cultivation, 49, 50, 63, 64, 69, 211n5
slavery, 83, 159
smallpox, 71
social movements, contemporary, 2, 203n3
social policy, and ethnography, 122–126
social practices/relations, changes in, 157–159, 162, 168–170, 200
social sciences, 2. *See also specific discipline*
South America, 173, 183, 184, 185. *See also specific country*
South East Wynad Planters Association, 83
South Kanara Catholic Association, 75–76
space, conceptualization of, 168
spice trade, 25
Spivak, Gayatri, 210n26
Spruce, Richard, 184, 187
Sri Lanka, 173
state control, over forests, 59–63
Stebbing, E. P., 20, 59
Stocking, George, 104–105, 114, 125, 220n35, 225n105
Stokes, Eric, 27, 226nn2, 4
subaltern studies, 42, 210n26
subordinated narratives, 42
Sullivan, John, 31–32
Swamis, 131
Syrian Christians, 160, 161, 218n18, 229n36

tahsildars (tax collectors), 50
takhal, 69, 70. *See also* slash-and-burn cultivation
taungya (planting system), 211n5
tea: diseases of, 217n34; plantations, 39
teak, 60, 63, 212n21
technological developments, in forestry, 76–77
technological knowledge, native, 113
technological modernity, 87–88, 152–155, 157–159. *See also* scientific modernity
technoscientific rationality, 9
Temple, R. C., 227n8
Temple, Richard, 151, 152, 227n8
Tentacles of Progress (Headrick), 87–89
terra nullius argument, 38
Third Decennial Congress of Protestant Missions, 157
Thompson, E. P., 52–53, 164, 230n43
Thomson, Thomas, 176
Thurston, Edgar, 14–15, 110–113, 119, 128, 129
timber felling, 67, 74
time, constructions of, 219n23
Todas (tribe), 14, 31, *32,* 99, 145–146, 208n6, 209n19, 219n21, 226n111; and biblical ethnology, 116–117; idolatry among, 45, 46, 220n43; lying by, 42–44; pedagogical efforts toward, 41–42; resistance of, 44; studies of, 109–110, 114, 219n26
Todas, The (Rivers), 110
trade, global, 24–25, 95–96. *See also* globalization/global trade
traditional cultures, preservation of, 102. *See also* premodern elements, retention of
Travancore, 159, 160, 166
Travels in Peru and India (Markham), 174, 182
Tribal Culture of India, The (Vidyarthi), 107–109
tribals, 13; assimilation of, 156–157; criminalization of, 22, 126–134, 162; education of, 67–69; English views of, 39, 40–44; and forest use, 102; as labor, 63–73, 82–84, 154, 162–163; and missionaries, 115, 159; as natural resources, 99; nose measurements of, 110, 111, *112*; photographs of, 221–222n58; and race, 141–142; relations between, 71–72, 145–146; resistance of, 64–65, 73–78; studies of, 108, 113, 124–125; as trackers, 18, 20; use of term, 204–205n14. *See also* indigenous people; natives
Tribes and Castes of Southern India (Thurston), 110–113, 119

United Planters Association of South India (UPASI), 83, 93, 94, 134, 207n40
upper-caste groups, 141, 152, 156, 160. *See also* Brahmins; elite, indigenous
Upper Godavari forests, 56
Ure, Andrew, 164
Utilitarians, 69, 150

Valmiki, 107–108, 131
Vedic *dasyus* (race), 225n102
venereal disease, among natives, 42
Victorian Anthropology (Stocking), 220n35
Victorian Britain, 26
Vidyarthi, L. P., 115, 121, 139–140
Viswanathan, Gauri, 151, 156, 227n7, 230n48, 235n8
"vocational" training, 68

Walker, Campbell, 189
Wallerstein, Immanuel, 1, 148
Western Ghats, 47, 51, 115
Western science, 5–6, 9–10. *See also* science
whiteness, "superiority" of, 143. *See also* race, classifications/concepts of; racial barriers
wildness, notions of, 17. *See also* nature

Williams, Raymond, 52–53, 157
Wolf, Eric, 12, 148
Wood, Sir Charles, 150, 227n7
work, 153–159; in missions, 153–154; and morality, 87, 154–156, 158–159, 215–216n11, 226n2; and progress, 152–155; social and political significance of, 156–157; and technological modernity, 157–159. *See also* labor
"working circles," 59
Wynad region, 47, 48, 63, 81

Yenadis (tribe), 99, 123–124, 125–126, 129
Yerukula (tribe), 229n39

About the Author

Kavita Philip received her doctorate in Science and Technology Studies from Cornell University, and is an associate professor at the University of California, Irvine. She was Senior Fellow 2002–2003 at the Rutgers Center for Historical Analysis. Her current research interests are in environmental history; postcolonial and feminist science studies; globalization, law, and human rights; and new media technologies. Her articles have appeared in the journals *Cultural Studies, Postmodern Culture, NMediaC, Irish Studies Review*, and *Environment and History*, and she serves on the editorial board of *Radical History Review.*